AF380549

Progress in Mathematics
Volume 249

Joachim Kock
Israel Vainsencher

An Invitation to Quantum Cohomology

Kontsevich's Formula for Rational Plane Curves

Birkhäuser
Boston • Basel • Berlin

Joachim Kock
Universitat Autònoma de Barcelona
Departament de Matemàtiques, Edifici C
08193 Bellaterra (Barcelona)
Spain

Israel Vainsencher
Universidade Federal de Minas Gerais
Departamento de Matemática — ICEx
30.123-970 Belo Horizonte — MG
Brazil

Mathematics Subject Classification (2000): 14N35, 14N10

Library of Congress Control Number: 2006924437

ISBN-10 0-8176-4456-3   e-ISBN: 0-8176-4495-4
ISBN-13 978-0-8176-4456-7

Printed on acid-free paper.

Birkhäuser 

Based on the original Portuguese edition, *A fórmula de Kontsevich para curvas racionais planas*, Instituto de Matemática Pura e Aplicada, Rio de Janeiro, Brazil, ©1999 J. Kock and I. Vainsencher

9 8 7 6 5 4 3 2 1

www.birkhauser.com   (TXQ/EB)

For Andréa and Kátia

Preface

This book is an elementary introduction to some ideas and techniques that have revolutionized enumerative geometry: *stable maps* and *quantum cohomology*. A striking demonstration of the potential of these techniques is provided by Kontsevich's famous formula, which solves a long-standing question:

> *How many plane rational curves of degree d pass through* $3d - 1$ *given points in general position?*

The formula expresses the number of curves for a given degree in terms of the numbers for lower degrees. A single initial datum is required for the recursion, namely, the case $d = 1$, which simply amounts to the fact that through two points there is but one line.

Assuming the existence of the Kontsevich spaces of stable maps and a few of their basic properties, we present a complete proof of the formula, and use the formula as a red thread in our *Invitation to Quantum Cohomology*. For more information about the mathematical content, see the Introduction.

The canonical reference for this topic is the already classical *Notes on Stable Maps and Quantum Cohomology* by Fulton and Pandharipande [29], cited henceforth as FP-NOTES. We have traded greater generality for the sake of introducing some simplifications. We have also chosen not to include the technical details of the construction of the moduli space, favoring the exposition with many examples and heuristic discussions.

We want to stress that this text is not intended as (and cannot be!) a substitute for FP-NOTES. *Au contraire*, we hope to motivate the reader to study the cited notes in depth. Have you got a copy? If not, point your browser to `http://arXiv.org/alg-geom/9608011`, and get it at once.

Prerequisites. We assume some basic algebraic geometry and some elementary intersection theory. For algebraic geometry: some familiarity with algebraic curves, divisors and line bundles, blowup, Grassmannians. Chapter 1 of Hartshorne [44] should be sufficient background, with some additional reading in Harris [42] for Grassmannians. We freely use the word "scheme" throughout, but do not make use of scheme theory in any essential way — in fact, we hardly use any commutative algebra. Spending an evening with Eisenbud–Harris [22] may be sufficient background on schemes. For intersection theory we just need the notions of pullback and pushforward of cycles and classes, the intersection product, the first Chern class of a line bundle, and Poincaré duality. The standard reference for this material is Fulton [28].

The original Portuguese edition of this book was written to support a five-lecture mini-course given at the 22° Colóquio Brasileiro de Matemática and published by IMPA in 1999. The idea of the mini-course and the style of the exposition go back to another mini-course, *Intersection theory over moduli spaces of curves* [33], taught by Letterio Gatto in the Recife Summer School, in January 1998. He showed us that it was possible to explain stable maps in an intelligible way, and that Kontsevich's formula was not just unattainable magic from theoretical physics: indeed it constitutes material that fits perfectly in the venerable tradition of enumerative geometry. The text for the mini-course grew gradually from notes from seminars given by the first author upon three occasions in 1998: in Recife, Belo Horizonte, and Maragogi, Alagoas.

This revised edition. Six years have past since the original Portuguese edition of this book appeared, and the subject of Gromov–Witten theory has evolved considerably. Speakers at conferences can nowadays say *stable map* with the same aplomb as six years ago they could say *stable curve*, assuming that the audience knows the definition, more or less. While the audience is getting used to the words, the magic surrounding the basics of the subject is still there, for better or for worse, both as fascinating mathematics, and sometimes as secret conjuration. However, for the student who wishes to get into the subject, the learning curve of FP-NOTES can appear quite steep. We feel there is still a need for a more elementary text on these matters, perhaps even more today, due to the rapid expansion of the subject. We hope this English translation can help in filling this gap.

This is a revised and expanded translation. Some errors have been corrected, some sections have been reorganized, and some clarifications of subtler points have been added. A short prologue with a few explicit statements on cross ratios and with a brief account of moduli spaces has been included; in Chapter 5 we have added a quick primer on generating functions. The five sections entitled "Generalizations and references" have been expanded, many more exercises have been included, and the Bibliography has been updated.

Acknowledgements. We thank the organizers of the 22° Colóquio Brasileiro de Matemática for the opportunity to give the minicourse, and the audience for their precious feedback. We are much indebted to Letterio Gatto for revealing the secrets of quantum cohomology to us, and for his support all along. Thanks are also due to Elizabeth Gasparim, Francesco Russo, and in particular Ty Le Tat, for reading and commenting on preliminary versions, and we are grateful to Steven Kleiman, Rahul Pandharipande, Aaron Bertram, Barbara Fantechi, and Anders Kock for kindly answering some questions related to the text. The first (resp. second) author was supported by a grant from the Danish Natural Science Research Council (resp. Brazil's CNPq) and registers here his gratitude.

Joachim Kock and *Israel Vainsencher*
Recife, April 1999
Barcelona & Belo Horizonte, August 2005

Contents

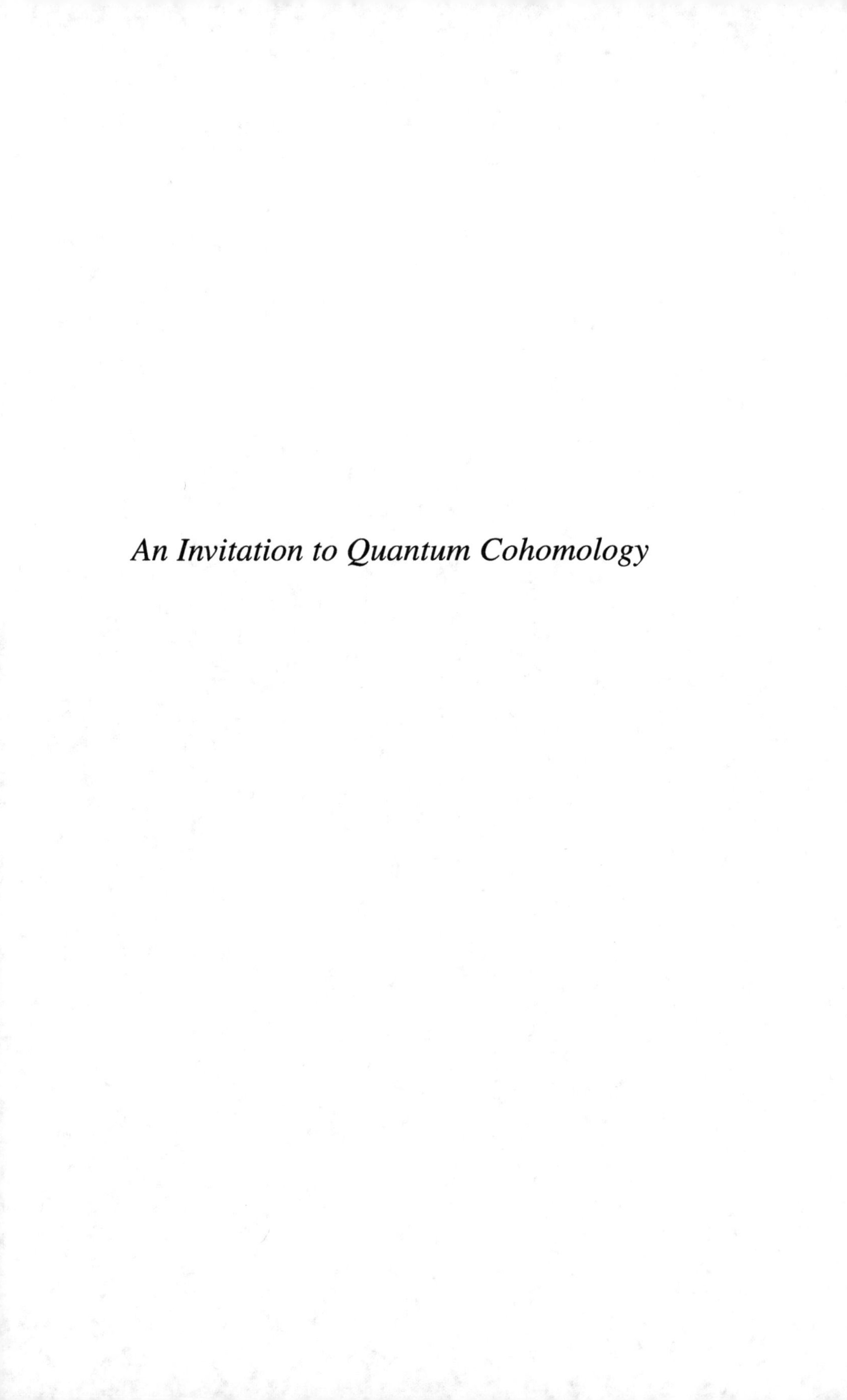

An Invitation to Quantum Cohomology

Introduction

What we are now witnessing on the geometry/physics frontier is, in my opinion, one of the most refreshing events in the mathematics of the 20th century. The ramifications are vast and the ultimate nature and scope of what is being developed can barely be glimpsed. It might well come to dominate the mathematics of the 21st century. ... For the students who are looking for a solid, safe PhD thesis, this field is hazardous, but for those who want excitement and action, it must be irresistible.

M. Atiyah [3]

The aim of enumerative geometry is to count how many geometric figures satisfy given conditions. The most basic example is the question, *How many lines are there through two distinct points?* A natural extension of this question is the problem of *determining the number N_d of rational curves of degree d passing through $3d-1$ points in general position in the complex projective plane.*[1] The number $3d-1$ is not arbitrary: it matches the dimension of the family of curves under consideration, so it is precisely the right number of conditions to impose in order to get a finite number of solutions.

The charm of such problems, which have enchanted mathematicians since the beginning of time, is that just as they are easy to state, the answer, if achieved, is of the simplest kind — after all it is but a natural number. Finding the solution, however, has often required quite innovative techniques. The numbers $N_1 = N_2 = 1$ go back to antiquity; $N_3 = 12$ was computed by Steiner [76] in 1848, but it was probably known earlier. The late 19th century was the golden era for enumerative geometry, and Zeuthen [87] computed the number $N_4 = 620$, as part of his comprehensive study of plane quartics. By then the art of resolving enumerative problems had attained a very high degree of sophistication, and in fact, its foundations could no longer support it. Included as the 15th problem in his list, Hilbert asked for a rigorous foundation of enumerative geometry. See Kleiman [50] for an interesting historic account with many references.

The 20th century witnessed great advances in intersection theory, an indispensable tool for enumerative geometry. In the seventies and eighties, many old

[1] Throughout we work over $\mathbb{C}$.

enumerative problems were solved, and many classical results were verified. However, the specific question of determining the numbers N_d turned out to be very difficult.

The revolution took place around 1994, when a connection between theoretical physics (string theory) and enumerative geometry was discovered. As a corollary, Kontsevich gave a solution to the old problem in terms of the recursive formula

$$N_d = \sum_{d_A+d_B=d} N_{d_A} N_{d_B} d_A^2 d_B \left(d_B \binom{3d-4}{3d_A-2} - d_A \binom{3d-4}{3d_A-1} \right).$$

Quite amazingly, it expresses the *associativity* of a certain new multiplication law, the *quantum product.* Not only does the formula allow the computation of as many of the numbers as you please, it also appeals to the aesthetic sensibility of mathematicians. After all, the determination of numbers that resisted a century of investigation is reduced to the number $N_1 = 1$ of lines passing through two distinct points! Furthermore, the formula appears as an instance of associativity, arguably one of the most basic concepts in mathematics.

The formula is a result of the new theories of *stable maps* and *quantum cohomology*, which have many other applications to enumerative geometry. Central objects in these theories are the moduli spaces of stable maps. The example treated in this book, $\overline{M}_{0,n}(\mathbb{P}^r, d)$, is a compactification of the space of isomorphism classes of maps $\mathbb{P}^1 \to \mathbb{P}^r$ of degree d, with n marked points, and subject to a certain stability condition. The numbers N_d occur as intersection numbers on the space $\overline{M}_{0,3d-1}(\mathbb{P}^2, d)$. In hindsight, these spaces are obvious parameter spaces for rational curves in $\mathbb{P}^2$: they are direct generalizations of the moduli spaces of stable curves, studied by Mumford and his collaborators [16] in the sixties and seventies.

However, historically, the path to Kontsevich's formula was a different one, and it is not quite wrong to say that the link to enumerative geometry came as a pleasant surprise. In string theory, developed by Witten and others (see for example [86]), the so-called topological quantum field theory introduced the notion of *quantum cohomology*. The terminology is due to Vafa [80]. Originally the coefficients for the quantum multiplication were described in terms of correlation functions, Lagrangians, and path integrals; the crucial discovery (mostly due to Witten) was that they could also be defined mathematically using algebraic or symplectic geometry. The relevant notion from symplectic geometry was Gromov's pseudoholomorphic curves [40]; hence the numbers were called *Gromov–Witten invariants*. The rigorous mathematical definition of these numbers required *moduli spaces of stable maps*. These were introduced independently by Ruan–Tian [73] in the symplectic category, and by Kontsevich–Manin [58] in the algebraic category.

Kontsevich and Manin set up axioms for Gromov–Witten invariants, and showed that these axioms imply the postulated properties of quantum cohomology, cf. Dubrovin [19]. In particular, they showed that the quantum product for $\mathbb{P}^2$ is associative if and only if the above recursive formula holds (as we shall see in Chapter 5). Once the formula was discovered, it was not difficult to provide a direct proof, which is basically the one we present in Chapter 3. The paper of Kontsevich and Manin [58] did not formally substantiate the existence of the moduli space $\overline{M}_{0,n}(\mathbb{P}^r, d)$, but it was constructed soon after by Behrend, Manin, and Fantechi (cf. [6], [8], [7]). The construction is quite technical and takes up almost 20 pages of FP-NOTES.

Let us describe briefly the contents of each chapter. The central notion of the subject is that of a *stable map*, introduced in Chapter 2. It is a natural extension of the notion of *stable curve*, the subject of Chapter 1.

To provide a smooth start to these notions, the Prologue offers a leisurely review of some easy facts about quadruples of points in $\mathbb{P}^1$ and cross ratios. These notions are fundamental for the subsequent material, and at the same time serve as a warm-up for the notion of moduli spaces reviewed in Section 0.2.

Chapter 1 is about *stable n-pointed curves* of genus 0 and their moduli space $\overline{M}_{0,n}$. This is the study of how rational curves break into *trees of rational curves*, and how n marked points can move on them, subject to a rule that whenever two marked points try to come together, a new component appears to separate them. The crucial feature of these spaces is the morphisms that connect them: by forgetting the last marked point one gets a morphism $\overline{M}_{0,n+1} \to \overline{M}_{0,n}$, which is the universal family, and by gluing together curves at marked points one gets morphisms $\overline{M}_{0,n+1} \times \overline{M}_{0,m+1} \to \overline{M}_{0,n+m+1}$ whose images form the locus of reducible curves. We describe $\overline{M}_{0,4}$ and $\overline{M}_{0,5}$ in considerable detail.

The second chapter constitutes the heart of the text: we define *stable maps* and describe their moduli spaces. We consider only stable maps of genus zero, and for target space we limit ourselves to $\mathbb{P}^r$; most of the time we stick to $\mathbb{P}^2$. We start with a heuristic discussion of maps $\mathbb{P}^1 \to \mathbb{P}^r$ and their degenerations, and motivate in this way the definition of stable map. We state without proof the theorem of existence of the coarse moduli spaces $\overline{M}_{0,n}(\mathbb{P}^r, d)$ of stable maps, and collect their basic properties: separatedness, projectivity, and normality, enough to allow us to do intersection theory on them.

We give only a very brief sketch of the idea behind the construction. Then we explore in more detail the important features of these spaces, many of which are inherited from the spaces $\overline{M}_{0,n}$: there are forgetful maps, which are almost universal families, and there are gluing maps, which produce reducible stable maps from irreducible ones and in this way give the moduli spaces a recursive structure; this is the key to Kontsevich's formula. At the end of the chapter we discuss in some more technical detail the folkloric comparison of $\overline{M}_{0,0}(\mathbb{P}^2, 2)$ with the classical space of complete conics.

In Chapter 3, we start out with a short introduction to the enumerative geometry of rational curves, comparing approaches based on *equations* (linear systems) with those based on *parametrizations* (maps). Next, we use the recursive structure of the space of stable maps to count conics (actually degree-2 stable maps) passing through 5 points. Then we move on to the counting of rational plane cubics (degree-3 stable maps) passing through 8 points. The arguments of these two examples are formalized to give a first proof of Kontsevich's formula. The noninterference of multiplicities is established via Kleiman's transversality theorem. We also check that counting maps is actually the same as counting curves.

In the last two chapters, we place Kontsevich's formula in its natural broader context, explaining the rudiments of *Gromov–Witten invariants* and *quantum cohomology*. In Chapter 4, we introduce Gromov–Witten invariants as a systematic way of organizing enumerative information, and we establish their basic properties. One crucial property is the Splitting Lemma, which expresses the easiest instance of the recursive structure of the moduli spaces. The examples of Chapter 3 are recast in this new language, so that Kontsevich's formula is subsumed as a particular case of the *Reconstruction Theorem*. This theorem states that all (genus-0) Gromov–Witten invariants of $\mathbb{P}^r$ can be computed from the first one, $I_1(h^r, h^r) = 1$, which is again nothing but the fact that through two distinct points there is a unique line!

The fifth and last chapter starts with a quick primer on generating functions; this formalism is ubiquitous in Gromov–Witten theory, but not an everyday tool for most algebraic geometers. We then define the Gromov–Witten potential as the generating function for the Gromov–Witten invariants, and use it to define the *quantum cohomology ring* of $\mathbb{P}^r$. We repeat in this new disguise the arguments of the preceding chapter in order to establish the *associativity* of this ring. Kontsevich's formula is now retrieved as a corollary of this property.

Throughout we have strived for simplicity, and as a consequence many results are not stated in their natural generality. Instead, each chapter ends with a section entitled *Generalizations and references*, where we briefly comment on generalizations of the material, provide a few historical comments, and outline topics that naturally extend the themes exposed in the text. We hope in this way to offer a guide for further reading. Some of the references are quite advanced, and certainly the present text does not contain sufficient background for their comprehension. Nevertheless, we recommend that the student skim through some of the papers to get a feeling for what is going on in this new and very active field, of which only some enumerative aspects are contemplated here.

There is a collection of exercises at the end of each chapter, ranging from routine verifications and hands-on computations to topics extending the main text, including a brief look at $\mathbb{P}^1 \times \mathbb{P}^1$ and $G(2, 4)$ as target spaces instead of $\mathbb{P}^r$. In particular, the exercises to Chapter 3 establish the $\mathbb{P}^1 \times \mathbb{P}^1$ version of Kontsevich's formula. The exercises are meant to be easy, and the student is encouraged to try to do them all.

Prologue: Warming Up with Cross Ratios, and the Definition of Moduli Space

Throughout this book we work over the field of complex numbers. When we speak of schemes we mean schemes of finite type over Spec $\mathbb{C}$.

0.1 Cross ratios

Let us warm up with a couple of reminders about simple facts concerning quadruples of points in $\mathbb{P}^1$, automorphisms of $\mathbb{P}^1$, and cross ratios. These notions are crucial to the subject of this book; besides, the space of cross ratios is one of the simplest nontrivial examples of a moduli space and serves as motivation for the general definition given in the next section.

0.1.1 Quadruples. By a *quadruple* of points in $\mathbb{P}^1$ we mean an ordered set of four distinct points $\mathbf{p} = (p_1, p_2, p_3, p_4)$ in $\mathbb{P}^1$. So the set of all quadruples forms an algebraic variety

$$Q := \mathbb{P}^1 \times \mathbb{P}^1 \times \mathbb{P}^1 \times \mathbb{P}^1 \smallsetminus \text{diagonals}.$$

We would like to say that Q is a fine moduli space for quadruples; that is, it should carry a universal family. We shall explain these terms as we go along and come back to their proper definition in the next section. Let us first specify what we mean by a family of quadruples.

0.1.2 Families of quadruples. A *family of quadruples* in $\mathbb{P}^1$ (over a base variety B) is a diagram

$$\begin{array}{c} B \times \mathbb{P}^1 \\ \pi \Big\downarrow \uparrow\uparrow\uparrow\uparrow \sigma_i \\ B \end{array} \qquad (\pi \circ \sigma_i = \mathrm{id}_B),$$

where π is the projection, and the four sections $\sigma_1, \ldots, \sigma_4 : B \to B \times \mathbb{P}^1$ are disjoint. So over each point $b \in B$, the fiber of π is a copy of the fixed $\mathbb{P}^1$, in which the sections single out four distinct points, hence a quadruple.

This formulation of the definition is made in order to resemble the notion of family we will use in Chapters 1 and 2. But for the situation at hand it is easy to see that the data is equivalent to simply giving a morphism

$$\sigma : B \to Q = \mathbb{P}^1 \times \mathbb{P}^1 \times \mathbb{P}^1 \times \mathbb{P}^1 \smallsetminus \text{diagonals.}$$

The disjointness of the sections $\sigma_1, \sigma_2, \sigma_3, \sigma_4$ is tantamount to the requirement that σ avoids the diagonals.

If $B \times \mathbb{P}^1 \to B$ (with its sections) is a family, and $\varphi : B' \to B$ is a morphism, then the pullback family is simply the family $B' \times \mathbb{P}^1 \to B'$ equipped with the four sections obtained by precomposing the four original sections with φ. This is better expressed in the second viewpoint: given a family $B \to Q$ and a morphism $B' \to B$, the pullback family is just the composite $B' \to B \to Q$.

0.1.3 The universal family over Q is now straightforward to describe: let the four sections σ_i be given by the four projections $Q \to \mathbb{P}^1$, $\sigma_i(\mathbf{p}) = (\mathbf{p}, p_i)$,

$$\begin{array}{c} Q \times \mathbb{P}^1 \\ \pi \big\downarrow \; \big\uparrow\big\uparrow\big\uparrow\big\uparrow \sigma_i \\ Q \end{array}$$

Translating into the second viewpoint, this is just the identity map $\mathrm{id}_Q : Q \to Q$. Clearly this family is *tautological* in the sense that the fiber over a point $\mathbf{p} \in Q$ is exactly the quadruple $\mathbf{p}$. But furthermore it enjoys a universal property: *every other family is induced from it by pullback.* This is particularly clear from the second viewpoint: every morphism $B \to Q$ factors through $\mathrm{id}_Q : Q \to Q$!

So Q classifies all quadruples. Things become more interesting when we ask for classification *up to projective equivalence*. That is, we now want an algebraic variety whose points are in natural bijection with the set of all equivalence classes of quadruples, and again we will require a universal property.

The notion of projective equivalence is defined in terms of automorphisms of $\mathbb{P}^1$.

0.1.4 Automorphisms of $\mathbb{P}^1$. The group of automorphisms of $\mathbb{P}^1$ is

$$\mathrm{Aut}(\mathbb{P}^1) \simeq \mathrm{PGL}(2),$$

the 3-dimensional group of invertible matrices $[\begin{smallmatrix} a & b \\ c & d \end{smallmatrix}]$ modulo a constant factor. It acts on $\mathbb{P}^1$ by multiplication, sending a point $[\begin{smallmatrix} x \\ y \end{smallmatrix}] \in \mathbb{P}^1$ to

$$\begin{bmatrix} a & b \\ c & d \end{bmatrix} \begin{bmatrix} x \\ y \end{bmatrix} = \begin{bmatrix} ax + by \\ cx + dy \end{bmatrix}.$$

In affine coordinates $x = [\begin{smallmatrix} x \\ 1 \end{smallmatrix}]$, the action takes the form of the familiar linear fractional transformation (also called a Möbius transformation) of one complex variable,

$$x \mapsto \frac{ax + b}{cx + d}.$$

0.1.5 Projective equivalence. Two quadruples $\mathbf{p}$ and $\mathbf{p}'$ are called *projectively equivalent* if there exists an automorphism $\phi : \mathbb{P}^1 \xrightarrow{\sim} \mathbb{P}^1$ such that $\phi(p_i) = p'_i$ for every $i = 1, 2, 3, 4$.

0.1.6 Moving points around in $\mathbb{P}^1$. A first question one could ask is whether all quadruples are projectively equivalent. This turns out not to be the case. However, up to three points can be moved around as we please: *Given any triple of distinct points $p_1, p_2, p_3 \in \mathbb{P}^1$, there exists a unique automorphism $\phi : \mathbb{P}^1 \xrightarrow{\sim} \mathbb{P}^1$ such that*

$$p_1 \mapsto 0, \quad p_2 \mapsto 1, \quad p_3 \mapsto \infty.$$

This construction defines a morphism

$$\alpha : \mathbb{P}^1 \times \mathbb{P}^1 \times \mathbb{P}^1 \smallsetminus \text{diagonals} \longrightarrow \operatorname{Aut}(\mathbb{P}^1).$$

You are asked to verify the details of these assertions in Exercise 1 on page 16.

0.1.7 The cross ratio. Let $\mathbf{p} = (p_1, p_2, p_3, p_4)$ be a quadruple. Let $\lambda(\mathbf{p}) \in \mathbb{P}^1$ denote the image of p_4 under the unique automorphism ϕ that sends p_1, p_2, p_3 to $0, 1, \infty$ (as above). The point $\lambda(\mathbf{p})$ is called the *cross ratio* of the quadruple $\mathbf{p}$.

Note that since the original four points were distinct, so are the four images $0, 1, \infty, \lambda(\mathbf{p})$; hence the cross ratio never attains any of the three values $0, 1, \infty$.

If all four points are distinct from ∞, then in affine coordinates $p_1 = [\begin{smallmatrix} x_1 \\ 1 \end{smallmatrix}]$, $p_2 = [\begin{smallmatrix} x_2 \\ 1 \end{smallmatrix}]$, $p_3 = [\begin{smallmatrix} x_3 \\ 1 \end{smallmatrix}]$, $p_4 = [\begin{smallmatrix} x_4 \\ 1 \end{smallmatrix}]$ the cross ratio is given by the formula

$$\lambda(p_1, p_2, p_3, p_4) = \frac{(x_2 - x_3)(x_4 - x_1)}{(x_2 - x_1)(x_4 - x_3)}.$$

This is the reason for the name cross ratio.

Note that the cross ratio map

$$\lambda : Q \longrightarrow \mathbb{P}^1 \smallsetminus \{0, 1, \infty\}$$

is a *morphism*. Indeed, it is the composition of these two morphisms:

$$\mathbb{P}^1 \times \mathbb{P}^1 \times \mathbb{P}^1 \times \mathbb{P}^1 \smallsetminus \text{diagonals} \xrightarrow{\alpha \times \mathrm{id}} \mathrm{Aut}(\mathbb{P}^1) \times \mathbb{P}^1 \longrightarrow \mathbb{P}^1,$$

where the second map is the action of $\mathrm{Aut}(\mathbb{P}^1)$ on $\mathbb{P}^1$, and α is as in 0.1.6.

0.1.8 Classification of quadruples up to projective equivalence. It is clear now that every quadruple $\mathbf{p}$ is projectively equivalent to $(0, 1, \infty, q)$ for a unique $q \in \mathbb{P}^1 \smallsetminus \{0, 1, \infty\}$, namely $q = \lambda(\mathbf{p})$. Therefore (by transitivity of the equivalence relation), *two quadruples are projectively equivalent if and only if they have the same cross ratio.* This shows that

> *the set $M_{0,4}$ of equivalence classes of quadruples is in a natural bijection with $\mathbb{P}^1 \smallsetminus \{0, 1, \infty\}$.*

In the symbol $M_{0,4}$, the index 4 refers of course to quadruples; the index 0 refers to the genus of the curve $\mathbb{P}^1$, anticipating the notation of Chapter 1.

Since $\mathbb{P}^1 \smallsetminus \{0, 1, \infty\}$ has the structure of an algebraic variety, we can carry that structure over to $M_{0,4}$. Now we will show that $M_{0,4}$ carries a universal family. This clarifies the technical meaning of naturality of the bijection.

0.1.9 The family over $M_{0,4}$. We will construct a tautological family of quadruples over $M_{0,4}$, that is, a family with the property that the fiber over any q is a quadruple with cross ratio q. The obvious choice for such a quadruple is $(0, 1, \infty, q)$, and this fits nicely into a family like this:

$$\begin{array}{c} U := M_{0,4} \times \mathbb{P}^1 \\ \pi \Big\downarrow \;\; \Big\uparrow\Big\uparrow\Big\uparrow\Big\uparrow \tau_i \\ M_{0,4} \end{array}$$

The first three sections are the constant ones equal to 0, 1, and ∞, and the last section is the "diagonal" $\delta : M_{0,4} \to M_{0,4} \times \mathbb{P}^1$ (via the inclusion $M_{0,4} \hookrightarrow \mathbb{P}^1$):

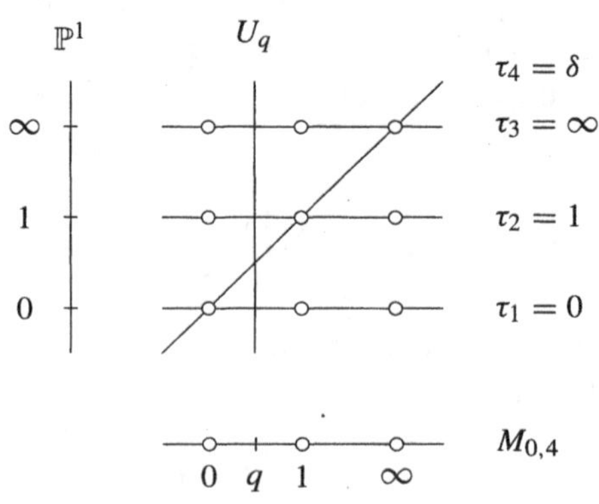

In the alternative formulation of 0.1.2, this family is given by

$$\begin{array}{rcl} \tau : M_{0,4} & \longrightarrow & Q \\ q & \longmapsto & (0, 1, \infty, q), \end{array}$$

and the tautological property translates into the observation that the composite

$$M_{0,4} \xrightarrow{\tau} Q \xrightarrow{\lambda} M_{0,4}$$

is the identity map.

Note that the other composite

$$Q \xrightarrow{\lambda} M_{0,4} \xrightarrow{\tau} Q,$$

takes any quadruple **p** to its "normal form" $(0, 1, \infty, q)$, where q is the cross ratio of **p**.

0.1.10 Projective equivalence in families. Two families $(B, \sigma_1, \sigma_2, \sigma_3, \sigma_4)$ and $(B, \sigma'_1, \sigma'_2, \sigma'_3, \sigma'_4)$ are equivalent if there is an automorphism $\phi : B \times \mathbb{P}^1 \xrightarrow{\sim} B \times \mathbb{P}^1$ making this diagram commute for $i = 1, \ldots, 4$:

$$\begin{array}{ccc} B \times \mathbb{P}^1 & \xrightarrow{\phi} & B \times \mathbb{P}^1 \\ \pi \downarrow \uparrow \sigma_i & & \pi' \downarrow \uparrow \sigma'_i \\ B & = & B \end{array}$$

This amounts to giving a morphism $B \to \mathrm{Aut}(\mathbb{P}^1)$, $b \mapsto \phi_b$, such that for each $b \in B$ the automorphism ϕ_b realizes an equivalence between the fibers over b (with the four sections).

In the second formulation, two families $\sigma : B \to Q$ and $\sigma' : B \to Q$ are projectively equivalent if there exists a morphism $\gamma : B \to \mathrm{Aut}(\mathbb{P}^1)$ such that this diagram commutes:

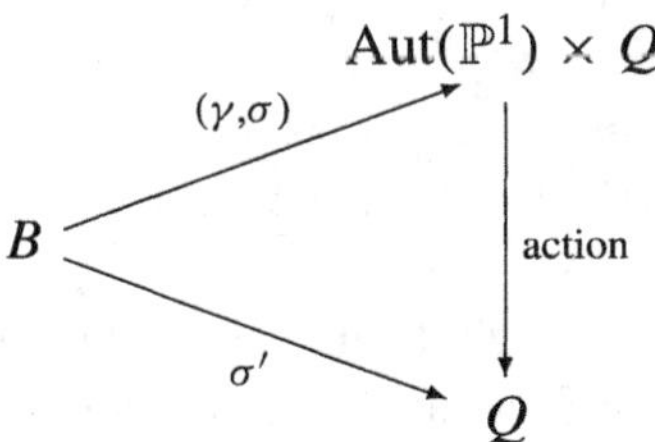

Just as two quadruples **p** and **p**$'$ are equivalent if and only if $\lambda(\mathbf{p}) = \lambda(\mathbf{p}')$, one can check that two families $\sigma : B \to Q$ and $\sigma' : B \to Q$ are equivalent if and only if $\lambda \circ \sigma = \lambda \circ \sigma'$. This requires some arguments analogous to the ones of the proof of the next lemma.

0.1.11 Lemma. *The tautological family $U \to M_{0,4}$ of 0.1.9 (with its four sections) has the universal property that for any other family $B \times \mathbb{P}^1 \to B$ (with its four sections) there is a unique morphism $\kappa : B \to M_{0,4}$, such that the family is projectively equivalent to the pullback along κ of $U \to M_{0,4}$ (with its four sections).*

Proof. Given a family $B \times \mathbb{P}^1 \to B$ (with its four sections), that is, a morphism $\sigma : B \to Q$, we just compose with $\lambda : Q \to M_{0,4}$ to get a morphism $\kappa : B \to M_{0,4}$. Now we must compute the pullback of the diagram

$$\begin{array}{ccc} & & M_{0,4} \times \mathbb{P}^1 \\ & & \pi \big\downarrow \;\; \big\uparrow\big\uparrow\big\uparrow\big\uparrow \sigma_i \\ B & \xrightarrow[\kappa]{} & M_{0,4} \end{array}$$

Since we are talking only about trivial families (in the sense that they are just products with $\mathbb{P}^1$), the only thing to worry about are the sections: the pulled-back sections are obtained by composition. This is easier to describe in the alternative viewpoint: the "pullback" of the family $M_{0,4} \xrightarrow{\tau} Q$ is simply the composite

$$B \xrightarrow{\kappa} M_{0,4} \xrightarrow{\tau} Q.$$

Now this family may not coincide with the original family $\sigma : B \to Q$, but we claim they are equivalent. Since $\kappa = \lambda \circ \sigma$, the picture is

$$\underbrace{\overbrace{B \xrightarrow{\sigma} Q}^{\text{original family}} \xrightarrow{\lambda} M_{0,4} \xrightarrow{\tau} Q}_{\text{pulled-back family}}$$

The map $Q \to M_{0,4} \to Q$ associates to each quadruple its "normal form" as we noted above. This map can also be described as the composite $Q \to \mathrm{Aut}(\mathbb{P}^1) \times Q \to Q$, where the first map is (α, id_Q) (taking the first three points to the unique automorphism as in 0.1.6), and the second is the action of $\mathrm{Aut}(\mathbb{P}^1)$ on Q. This gives a morphism $B \to Q \to \mathrm{Aut}(\mathbb{P}^1) \times Q$, which is just the one needed to see that the two families are equivalent.

It remains to check that the map κ is unique with the required pullback property. But in order for the pullback of U along κ to be equivalent to the original family, κ must have the property that the image of a point b is the cross ratio of the fiber over b. Clearly this determines κ completely, if it exists. □

The existence of a family with that universal property amounts to the following result:

0.1.12 Proposition. *$M_{0,4}$ is a fine moduli space for the problem of classifying quadruples in $\mathbb{P}^1$ up to projective equivalence.*

The definition of fine moduli space will be given in a moment.

0.1.13 n-tuples. More generally, we can classify n-tuples up to projective equivalence: two n-tuples

$$(p_1, p_2, p_3, \dots, p_n) \text{ and } (p'_1, p'_2, p'_3, \dots, p'_n)$$

are projectively equivalent in $\mathbb{P}^1$ if and only if the identity of cross ratios

$$\lambda(p_1, p_2, p_3, p_i) = \lambda(p'_1, p'_2, p'_3, p'_i)$$

holds for each $i = 4, \dots, n$. Now this classification problem is solved by the fine moduli space

$$M_{0,n} \simeq M_{0,4} \times \cdots \times M_{0,4} \smallsetminus \text{diagonals } (n-3 \text{ factors}),$$

the universal family being given by

$$\begin{array}{c} M_{0,n} \times \mathbb{P}^1 \\ \pi \big\downarrow \big\uparrow \sigma \\ M_{0,n} \end{array}$$

where the sections are the projections from the $(n-3)$-fold product to its factors $M_{0,4} \subset \mathbb{P}^1$.

We will come back to this in Chapter 1 in a slightly more general setting.

0.2 Definition of moduli space

0.2.1 Moduli spaces. Roughly speaking, a moduli problem consists in classifying certain geometric objects up to a given notion of equivalence. The objects could be varieties or schemes, vector bundles, maps, etc., of a specified type. The notion of equivalence could be projective equivalence, isomorphism, etc. The desired *moduli space* ought to be a variety or scheme, whose geometrical points are in natural bijection with the set of equivalence classes of the objects, and satisfying some further properties, which we proceed to make precise. The very meaning of "natural" required for the bijection is at the heart of things here.

We first treat the notion of fine moduli space (of which the space of cross ratios is an example), which is conceptually straightforward. Then we come to the notion of coarse moduli space, which is a bit subtler, although much more common in nature.

0.2.2 Notions of families and equivalence. It is not sufficient to have a bijective correspondence between the equivalence classes of objects and the points of the moduli space M: furthermore we want the algebraic structure of M to capture and reflect the way the objects vary in families. So the formulation of a moduli problem begins with a notion of a family of objects over a base scheme B, together with a notion of pullback of families along morphisms. This means that if $\mathfrak{X}$ is a family over B, which we denote by writing $\mathfrak{X}/B$, and we have a morphism $\varphi : B' \to B$, then there is induced a family over B', denoted by $\varphi^*\mathfrak{X}/B'$.

Often (and in all examples we will be concerned with), the notion of family will consist of a morphism $\mathfrak{X} \to B$ equipped with some extra structure, and then the members of the family are the fibers equipped with the extra structure induced from $\mathfrak{X}$. For example, we had the notion of family of quadruples given by a map $B \times \mathbb{P}^1 \to B$ where the extra structure consists of certain sections, and in Chapter 2 we will have families of maps, which are diagrams like

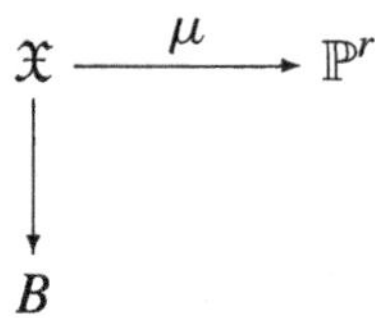

where the additional structure is the map μ. Often some further condition must be imposed on $\mathfrak{X} \to B$: almost always it is necessary to impose a flatness condition.

In all these examples, the pullback operation is nothing but the usual fiber product, $\varphi^*\mathfrak{X} := B' \times_B \mathfrak{X}$, with the first projection as structure map. In each case you should make explicit how the extra structure is induced on the pullback family. The usual pullback enjoys two easy properties (which should be imposed as axioms for an abstract pullback operation): pulling back along the identity morphism gives back the same family again, and pulling back along a composite $\psi \circ \varphi$ is the same as pulling back first along φ then along ψ.

Next, there should be a notion of equivalence of families compatible with the pullback operation in this sense: if $\mathfrak{X}$ and $\mathfrak{Y}$ are equivalent families over B (this will be denoted by $\mathfrak{X} \simeq \mathfrak{Y}$ in this section), and if $\varphi : B' \to B$ is any morphism, then $\varphi^*\mathfrak{X} \simeq \varphi^*\mathfrak{Y}$ as families over B'.

In fact, since the aim is classification of objects up to equivalence, it is acceptable if the notion of pullback is defined only up to equivalence, in which case the two axioms hold only up to equivalence too.[2]

[2]This is actually the usual case, although such subtleties are often suppressed: the fiber product of two varieties or schemes is really defined only up to (unique) isomorphism; it is possible to make choices once and for all to make it well-defined, but there is no canonical choice and the price to pay is that the composition axiom and the identity arrow axiom will hold only up to isomorphism.

In all cases of interest, where the family is given in terms of the fibers of a morphism, the set of all equivalence classes of families over the point $\bullet := \operatorname{Spec} \mathbb{C}$ is in natural bijection with the set of all equivalence classes of objects.

0.2.3 Fine moduli spaces. A *universal family* for a moduli problem is a family U/M with the property that for any family $\mathfrak{X}/B$ there exists a unique morphism $\kappa : B \to M$ such that κ^*U is equivalent to $\mathfrak{X}$ as families over B. The base of a universal family is called a *fine moduli space.*

In other words, for each base B there is a bijection between the set of families over B (up to equivalence) and the set of morphisms $B \to M$.

It is a consequence of the universal property that a universal family for a given moduli problem is unique up to equivalence, if it exists (Exercise 10 on page 18).

0.2.4 Points of the moduli space. The formulation of fine moduli space is all about families, but in the special case $B = \bullet$ we get the classification of objects: a family over $\bullet$ is just an object, and a morphism $\bullet \to M$ is a geometric point of M. So the geometric points of M are in bijective correspondence with equivalence classes of objects, which was the original goal.

The morphism $\kappa : B \to M$ corresponding to a family $\mathfrak{X}/B$ is called the *classifying map* of the family: it sends each point $b \in B$ to the point in M classifying the fiber $\mathfrak{X}_b$, i.e., the unique point in M whose fiber is equivalent to $\mathfrak{X}_b$.

Tautologically, the fiber U_m over a moduli point $m \in M$ belongs to the equivalence class corresponding to m under the bijection. A family with this property is called *tautological.* Not every tautological family is universal.

0.2.5 Toward a categorical reformulation. The definition of fine moduli space can be reformulated in terms of representable functors, as we proceed to do. This formulation requires a few categorical concepts and results not really used in the remaining chapters of these notes, but it is a very rewarding abstraction; virtually every universal property in mathematics can be formulated in terms of representable functors (cf. [60], Ch. 3). In any case, the treatment of coarse moduli spaces becomes much neater with the categorical language.

0.2.6 The moduli functor. The conditions imposed on the notions of family, pull-back, and equivalence amount precisely to saying that we have a functor

$$\begin{aligned} F : \boldsymbol{Sch}^{\mathrm{op}} &\longrightarrow \boldsymbol{Set} \\ B &\longmapsto \{\text{equivalence classes of families over } B\}. \end{aligned}$$

Here $\boldsymbol{Sch}^{\mathrm{op}}$ denotes the opposite category of the category of schemes, which is just a way of saying that F is contravariant, i.e., reverses the direction of arrows. The display tells only what F does on objects; it is equally important to tell what it does on arrows: it sends a morphism $\varphi : B' \to B$ to the pullback map $\varphi^* : F(B) \to F(B')$. Thus, given a family $\mathfrak{X}/B$ (hence representing an element in the set $F(B)$) we associate the family $\varphi^*\mathfrak{X}/B'$ (representing an element in the set $F(B')$). The conditions imposed on pullback amount to the axioms for a functor, namely to respect composition of arrows and identity arrows. The compatibility with equivalence means that pullback is well defined on equivalence classes, not just on actual families.

In conclusion, a moduli problem is concisely encoded as a contravariant functor from schemes to sets.

0.2.7 Representable functors. Any scheme Y gives rise to a contravariant set-valued functor called its *functor of points*,

$$\begin{aligned} \mathrm{h}_Y : \boldsymbol{Sch}^{\mathrm{op}} &\longrightarrow \boldsymbol{Set} \\ B &\longmapsto \mathrm{Hom}(B, Y). \end{aligned}$$

Here $\mathrm{Hom}(B, Y)$ denotes the set of all morphisms $B \to M$. A morphism $\varphi : B' \to B$ is sent to the map of sets

$$\begin{aligned} \mathrm{Hom}(B, Y) &\longrightarrow \mathrm{Hom}(B', Y) \\ \beta &\longmapsto \beta \circ \varphi. \end{aligned}$$

A functor F isomorphic to a functor of this sort is called *representable*, and if $\mathsf{u} : \mathrm{h}_Y \xrightarrow{\sim} F$ is an isomorphism of functors we say that the pair (Y, u) *represents F*.

By Yoneda's lemma (Exercise 11 on page 18), the natural transformation $\mathsf{u} : \mathrm{h}_Y \to F$ is identified with an element U in the set $F(Y)$; we will also say that the pair (M, U) represents F. Now we are ready for the categorical reformulation: *a family U/M is a universal family (and M a fine moduli space) for F if and only if the pair (M, U) represents F* (Exercise 12 on page 19).

0.2.8 Example: Hilbert schemes. Important examples of fine moduli spaces are Hilbert schemes. These represent moduli functors of flat families of closed subschemes in $\mathbb{P}^r$ with given Hilbert polynomial. The equivalence relation is just equality. The universal family U/M is parametrized by a projective scheme M which is known to be connected but, in general, not reduced or irreducible. The total space of the family is a closed subscheme $U \subseteq M \times \mathbb{P}^r$. Each geometric fiber U_m gives a closed subscheme of $\mathbb{P}^r$ with the given Hilbert polynomial and any such subscheme occurs for precisely one $m \in M$. See the notes of Strømme [78] for an elementary introduction to Hilbert schemes.

For many interesting moduli functors, a fine moduli space does not exist, and we must look for a weaker solution to the classification problem: a *coarse moduli space*. The idea is to relax the condition a little bit: we still want the geometric points of the moduli space to be in bijection with the equivalence classes of objects, but if the moduli functor is not representable we will look instead for sort of a best possible representable approximation, a scheme satisfying a weaker universal property.

The central moduli space of this book, the space of stable maps introduced in Chapter 2 is such a coarse moduli space.

Nonexistence of fine moduli spaces is often related to the presence of nontrivial automorphisms of the objects, but as illustrated by Exercise 15 on page 19, this is not a formal obstruction.

0.2.9 Coarse moduli spaces. A *coarse moduli space* for a moduli functor F is a pair (M, v), where M is a scheme and $\mathsf{v} : F \to \mathrm{h}_M$ is a natural transformation (not required to be invertible), such that

(i) (M, v) is *initial* among all such pairs.
(ii) The set map $\mathsf{v}_\bullet : F(\bullet) \to \mathrm{Hom}(\bullet, M)$ is a bijection.

That (M, v) is initial means the following: given any other pair (M', v') where $\mathsf{v}' : F \to \mathrm{h}_{M'}$ is a natural transformation, then there exists a unique morphism $\psi : M \to M'$ such that $\mathsf{v}' = \psi \circ \mathsf{v}$. By Yoneda's lemma (Exercise 11 on page 18), natural transformations $\mathrm{h}_M \to \mathrm{h}_{M'}$ correspond bijectively to morphisms $M \to M'$, so by abuse of notation we also use the symbol ψ for the corresponding natural transformation $\mathrm{h}_M \to \mathrm{h}_{M'}$. In other words, every natural transformation $F \to \mathrm{h}_{M'}$ factors uniquely through v, and in this sense h_M is the representable functor closest to F.

It is not difficult to check (Exercise 13 on page 19) that a pair (M, v) satisfying (i) is unique (up to unique comparison isomorphism) if it exists. Note that a fine moduli space is also a coarse moduli space (take $\mathsf{v} = \mathsf{u}^{-1}$, with the above notations). By uniqueness, it follows that if a coarse moduli space exists and if it is not fine, then a fine moduli space cannot exist.

Condition (ii) says the geometric points of M parametrize the equivalence classes of objects. We shall see in a moment that this condition is not automatic: there are situations in which a pair (M, v) exists that satisfies (i) but not (ii).

0.2.10 Example of nonexistence of moduli. To finish this section, we describe a situation in which moduli spaces cannot exist, as an illustration of violated continuity.

For a given moduli problem, suppose there exists a family $\mathfrak{X}/B$ over some irreducible variety B such that all of its members are equivalent except one special member, which is of another class. Then a coarse moduli space can not exist.

Indeed, suppose a coarse moduli space M existed. The family $\mathfrak{X}/B$ would induce a classifying map $B \to M$, but this morphism would map all but a point of B to one point of M, and the special point to a different one; hence the classifying map would lack the continuity necessary for being a morphism.

There are lots of natural examples of this situation: the moduli problem of "two points in $\mathbb{P}^1$, possibly coincident, up to projective equivalence." The moduli problem of "trees of projective lines up to combinatorial type," the moduli problem of "conics up to isomorphism." In each of these examples, it follows readily that $\bullet = \operatorname{Spec}\mathbb{C}$ equipped with the trivial natural transformation $F \to \mathrm{h}_\bullet$ satisfies condition (i) of the definition of coarse moduli space. Hence condition (ii) is not an automatic consequence of condition (i).

Exercises

1. (i) Show that any triple $\mathbf{p} = (p_1, p_2, p_3)$ of distinct points in $\mathbb{P}^1$ is projectively equivalent to the standard triple $(0, 1, \infty)$ by a unique automorphism $\phi_{\mathrm{p}} \in \operatorname{Aut}(\mathbb{P}^1)$.

   (ii) Write down the 2×2 matrix of ϕ_{p} in terms of the trihomogeneous coordinates of $\mathbf{p}$ and show that the map
   $$\alpha : \mathbb{P}^1 \times \mathbb{P}^1 \times \mathbb{P}^1 \smallsetminus \text{diagonals} \longrightarrow \operatorname{Aut}(\mathbb{P}^1)$$
   is a morphism.

2. Show that any quadruple of points in $\mathbb{P}^2$ such that no three lie on a line is projectively equivalent to the coordinate points
   $$\begin{bmatrix}1\\0\\0\end{bmatrix}, \begin{bmatrix}0\\1\\0\end{bmatrix}, \begin{bmatrix}0\\0\\1\end{bmatrix}, \begin{bmatrix}1\\1\\1\end{bmatrix}.$$

3. We say n points in $\mathbb{P}^r = \mathbb{P}(V)$ form an *independent set* if they are represented by independent vectors in the underlying $(r+1)$-dimensional vector space V. We say m points in $\mathbb{P}^r$ are in *general position* if any subset of $n \leq r+1$ points is an independent set. Note that for points in $\mathbb{P}^1$, general position simply means that no two of them coincide.

   Generalizing the previous two exercises, show that every $(r+2)$-tuple of points in general position in $\mathbb{P}^r$ is projectively equivalent to the $(r+2)$-tuple of coordinate points
   $$\begin{bmatrix}1\\0\\\vdots\\0\end{bmatrix}, \begin{bmatrix}0\\1\\\vdots\\0\end{bmatrix}, \ldots, \begin{bmatrix}0\\\vdots\\0\\1\end{bmatrix}, \begin{bmatrix}1\\1\\\vdots\\1\end{bmatrix}.$$

Unordered quadruples and the j-invariant

4. The symmetric group on four letters $\mathfrak{S}_4$ acts on $Q = \mathbb{P}^1 \times \mathbb{P}^1 \times \mathbb{P}^1 \times \mathbb{P}^1 \smallsetminus$ diagonals by permutation of the factors.

   (i) Show that this action descends to an action on $M_{0,4} \simeq \mathbb{P}^1 \smallsetminus \{0, 1, \infty\}$.

   (ii) Show that if the permutation σ consists of two disjoint 2-cycles, then the cross ratio of $\sigma(\lambda)$ is again λ, for every $\lambda \in M_{0,4}$.

   (iii) Show that these double 2-cycles span a normal subgroup of $\mathfrak{S}_4$ of order 4, and that the quotient is naturally isomorphic to $\mathfrak{S}_3$, the permutation group of $\{0, 1, \infty\}$.

5. (i) Show that the $\mathfrak{S}_3$-orbit of λ is the following set of cross ratios,

$$\Lambda(\lambda) = \left\{\lambda, \quad 1-\lambda, \quad \lambda^{-1}, \quad \frac{\lambda}{\lambda-1}, \quad \frac{1}{1-\lambda}, \quad \frac{\lambda-1}{\lambda}\right\},$$

   corresponding to the permutations Id, (01), (0∞), (1∞), (01∞), $(0\infty1)$.

   (ii) Show that two unordered quadruples of points in $\mathbb{P}^1$ are congruent (i.e., there is an automorphism taking one set to the other, irrespective of the orderings) if and only if their $\mathfrak{S}_3$-orbits coincide.

6. There are some cross ratios more special than others:

   (i) Show that the three involutions (01), (0∞), and (1∞) each has a fixed point (respectively, $\frac{1}{2}$, -1, and 2), and that each of the 3-cycles (01∞) and $(0\infty1)$ has two fixed points, $\xi_\pm := \frac{1\pm\sqrt{-3}}{2}$.

   (ii) Show that the involutions interchange $\xi_\pm$; they also interchange the two remaining special cross ratios. The 3-cycles permute the points $\frac{1}{2}$, -1, 2.

7. The j-invariant of a quadruple of points in $\mathbb{P}^1$ is by definition

$$j(\lambda) \;=\; 2^8\,\frac{(\lambda^2-\lambda+1)^3}{\lambda^2(1-\lambda)^2},$$

   where λ is the cross ratio of the quadruple. (The factor 2^8 comes from number theory but it plays no role in the present context.) Show that $j(\lambda) = j(\lambda')$ if and only if $\Lambda(\lambda) = \Lambda(\lambda')$ (notation as above). Conclude that two unordered quadruples are congruent if and only if they have the same j-invariant.

   Note that the points $\xi_\pm$ have j-invariant 0, and that the points $\frac{1}{2}$, -1, and 2 have j-invariant $1728 = 2^6 \cdot 3^3$.

8. Usually j-invariants are associated to elliptic curves. Given an elliptic curve $Y^2 = X(X - Z)(X - \lambda Z)$, its projection onto the line $Y = 0$ is ramified over the four points $0, 1, \infty, \lambda$. Show that two such curves are projectively equivalent if and only if they share the same set $\Lambda(\lambda)$. *Hint*: first show that if P, Q are flexes of a smooth plane cubic C, then there exists $\phi \in \mathrm{Aut}(\mathbb{P}^2)$ such that $\phi(C) = C, \phi(P) = Q$.

9. Show that the affine line (the values of the j-invariant) is a coarse moduli space for unordered quadruples up to congruence. To make sense of this, we first need a notion of family of unordered quadruples: it is a flat morphism

$$\begin{array}{c} \mathfrak{X} \subset B \times \mathbb{P}^1 \\ \big\downarrow \\ B \end{array}$$

such that every fiber consists of four distinct points. The universal family is parametrized by the open subset of $\mathbb{P}^4$ of binary quartics with nonzero discriminant. You can try to write down a formula for j in terms of the coefficients of the quartic.

Representable functors and universal families

10. Show that a universal family for a moduli problem is unique up to equivalence, if it exists.

11. *Yoneda Lemma.* Let Y be an object in an arbitrary category $\boldsymbol{S}$, let $\mathrm{h}_Y : \boldsymbol{S}^{\mathrm{op}} \to \boldsymbol{Set}$ denote the functor $X \mapsto \mathrm{Hom}(X, Y)$, and let $F : \boldsymbol{S}^{\mathrm{op}} \to \boldsymbol{Set}$ be any functor. Recall that a natural transformation $\mathsf{u} : \mathrm{h}_Y \to F$ consists of a map of sets $\mathsf{u}_X : \mathrm{h}_Y(X) = \mathrm{Hom}(X, Y) \to F(X)$ for each object X, compatible with morphisms in the sense that for every morphism $\varphi : X' \to X$ this diagram commutes:

$$\begin{array}{ccc} \mathrm{Hom}(X, Y) & \xrightarrow{\mathsf{u}_X} & F(X) \\ {\scriptstyle \_\circ\varphi}\big\downarrow & & \big\downarrow{\scriptstyle F(\varphi)} \\ \mathrm{Hom}(X', Y) & \xrightarrow[\mathsf{u}_{X'}]{} & F(X') \end{array}$$

We write $\mathrm{Nat}(\mathrm{h}_Y, F)$ for the set of natural transformations from h_Y to F.

Prove the Yoneda Lemma: *There is a natural bijection* $\mathrm{Nat}(\mathrm{h}_Y, F) \leftrightarrow F(Y)$.

Hint: Given a natural transformation $\mathsf{u} : \mathrm{h}_Y \to F$, the corresponding element in $F(Y)$ is $u := \mathsf{u}_Y(\mathrm{id}_Y)$. Use naturality to show that u is uniquely determined by u via the formula $\mathsf{u}_X(\alpha) = F(\alpha)(u)$. Conversely, for an arbitrary element $u \in F(Y)$, use this formula to define a natural transformation $\mathsf{u} : \mathrm{h}_Y \to F$.

12. Show that a family U/M is a universal family if and only if the pair (M, U) represents the moduli functor. *Hint:* if U/M is a universal family, naturality of the bijections $\mathrm{Hom}(B, M) \to F(B)$, $\kappa \mapsto \kappa^*U$ follows from the equation $(\kappa \circ \varphi)^*U \simeq \varphi^*(\kappa^*U)$ (for any morphism $\varphi : B' \to B$).

13. Show that a coarse moduli space is unique up to unique isomorphism, if it exists. In fact, already condition (i) in 0.2.9 is enough to guarantee this.

14. *The subset classifier.* Consider the moduli problem of classifying subsets. A family over a set B is defined to be a subset of B, and two subsets are considered equivalent when equal. Show that the inclusion

$$\begin{array}{c}\{\texttt{true}\}\\ \cap\\ \{\texttt{true}, \texttt{false}\}\end{array}$$

is a universal family.

15. *Classifying finite sets.* Counting is one of the most fundamental aspects of mathematics, and in this book we are very much concerned with counting. Counting makes sense because the following moduli problem has a fine solution: the problem is to classify all finite sets up to isomorphism. The solution must be a set whose elements correspond to the isomorphism classes of finite sets. It was one of the first abstract mathematical achievements of mankind to invent the natural numbers for the purpose of classifying finite sets. (Of course this invention came gradually, starting perhaps with the set $\{1, 2, 3, 4, 5\}$ classifying nonempty finite sets with at most five elements, and so on, until at some point the huge step of abstraction was taken and the set $\mathbb{N}$ was invented. Different cultures did not use exactly the same classifying sets, illustrating the fact that a fine moduli space is only determined up to isomorphism.)

A *family of finite sets* is just a set map $p : E \to B$ such that each fiber $p^{-1}(b)$ is a finite set. Given a map of sets $\varphi : B' \to B$, the pullback $\varphi^*E := B' \times_B E$ is a family of finite sets over B'. Declare two families equivalent if there is a fiberwise bijection between them. Show that $\mathbb{N}$ is a fine moduli space for this moduli problem by giving a concrete description of the universal family. Describe the classifying map. Restate the arguments in categorical terms.

Notice that finite sets of cardinality greater than 1 have nontrivial automorphisms. Hence we see by way of example that automorphisms do not constitute a formal obstruction to the existence of a fine moduli space.

Chapter 1

Stable n-pointed Curves

Our main object of study will be the spaces $\overline{M}_{0,n}(\mathbb{P}^r, d)$ of stable maps, whose introduction we defer to Section 2.3. Many of the properties of these spaces are inherited from $\overline{M}_{0,n}$, the important Deligne–Mumford–Knudsen moduli space of stable n-pointed rational curves which are the subject of this first chapter. We shall not go into the detail of the construction of $\overline{M}_{0,n}$, but content ourselves with the cases $n \leq 5$. The combinatorics of the boundary deserves a careful description. The principal reference for this chapter is Knudsen [51]; see also Keel [47].

1.1 n-pointed smooth rational curves

Definition. An *n-pointed smooth rational curve*

$$(C, p_1, \dots, p_n)$$

is a projective smooth rational curve C equipped with a choice of n distinct *marked points* $p_1, \dots, p_n \in C$.

An isomorphism between two n-pointed rational curves

$$\varphi : (C, p_1, \dots, p_n) \xrightarrow{\sim} (C', p'_1, \dots, p'_n)$$

is an isomorphism $\varphi : C \xrightarrow{\sim} C'$ that respects the marked points (in the given order), i.e.,

$$\varphi(p_i) = p'_i, \quad i = 1, \dots, n.$$

More generally, a *family of n-pointed smooth rational curves* is a flat and proper map (cf. [44, p. 95, 253]) $\pi : \mathfrak{X} \to B$ with n disjoint sections $\sigma_i : B \to \mathfrak{X}$ such that each geometric fiber $\mathfrak{X}_b := \pi^{-1}(b)$ is a projective smooth rational curve. Note that the n sections single out n special points $\sigma_i(b)$ which are the n marked points of that fiber. An *isomorphism* between two families $\pi : \mathfrak{X} \to B$ and $\pi' : \mathfrak{X}' \to B$ (with the same base) is an isomorphism $\varphi : \mathfrak{X} \xrightarrow{\sim} \mathfrak{X}'$ making this diagram commutative:

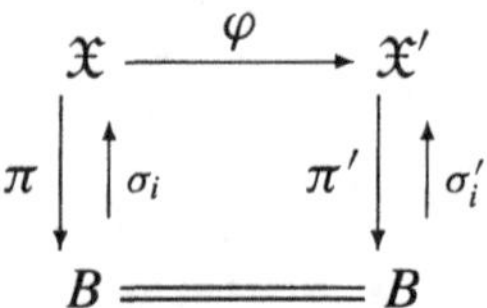

1.1.1 Comparison with n-tuples of points on a fixed $\mathbb{P}^1$. If $\mathfrak{X} \to B$ is a flat family with geometric fibers isomorphic to $\mathbb{P}^1$ that admits at least one section, one can show that $\mathfrak{X} \simeq \mathbb{P}(\mathcal{E})$ for some rank-2 vector bundle $\mathcal{E}$ on B (cf. the argument in [44, p. 369]). If the family admits at least two disjoint sections, then the bundle splits; and if there are at least three disjoint sections, one can show that $\mathfrak{X} \simeq B \times \mathbb{P}^1$ and that there is a unique isomorphism such that the three sections are identified with the constant sections $B \times \{0\}$, $B \times \{1\}$, $B \times \{\infty\}$, in this order. Thus, if $n \geq 3$, for any given family $\mathfrak{X} \to B$ of n-pointed smooth curves there is a unique B-isomorphism $\mathfrak{X} \to B \times \mathbb{P}^1$; so all the information is in the sections. In other words, for $n \geq 3$, to classify n-pointed smooth rational curves up to isomorphism is the same as classifying n-tuples of distinct points in a fixed $\mathbb{P}^1$, up to projective equivalence, as in the Prologue.

So the following result is just a reformulation of 0.1.13.

1.1.2 Proposition. *For $n \geq 3$, there is a fine moduli space, denoted by $M_{0,n}$, for the problem of classifying n-pointed smooth rational curves up to isomorphism.*

This means that there exists a universal family $U_{0,n} \to M_{0,n}$ of n-pointed curves. Thus, every family $\mathfrak{X} \to B$ of projective smooth rational curves equipped with n disjoint sections is induced (together with the sections) by pullback along a unique morphism $B \to M_{0,n}$. (Again, the index 0 in the symbol $M_{0,n}$ refers to the genus of the curves.)

1.1.3 Example. Let $n = 3$. Given any smooth rational curve with three marked points (C, p_1, p_2, p_3), there is a unique isomorphism to $(\mathbb{P}^1, 0, 1, \infty)$. That is, there exists only one isomorphism class, and consequently, $M_{0,3}$ is a single point. Its universal family is $\mathbb{P}^1$ with marked points $0, 1, \infty$.

1.1.4 Example. Suppose now $n = 4$. This is the first nontrivial example of a moduli space of pointed rational curves, and we already described it carefully in 0.1.9. Every rational curve with four distinct marked points, (C, p_1, p_2, p_3, p_4), is isomorphic to $(\mathbb{P}^1, 0, 1, \infty, q)$ for some unique $q \in \mathbb{P}^1 \smallsetminus \{0, 1, \infty\}$, and the isomorphism of pointed curves is unique. Thus our space of cross ratios $M_{0,4} \simeq \mathbb{P}^1 \smallsetminus \{0, 1, \infty\}$ is the moduli space for 4-pointed smooth rational curves (up to isomorphism). The universal family is the trivial family $U_{0,4} := M_{0,4} \times \mathbb{P}^1 \to M_{0,4}$ equipped with the following disjoint sections: the three constant sections $\tau_1 = M_{0,4} \times \{0\}$, $\tau_2 = M_{0,4} \times \{1\}$, and $\tau_3 = M_{0,4} \times \{\infty\}$, together with the diagonal section τ_4. Now the fiber over a point $q \in M_{0,4}$ is a projective line U_q with four points singled out by the sections

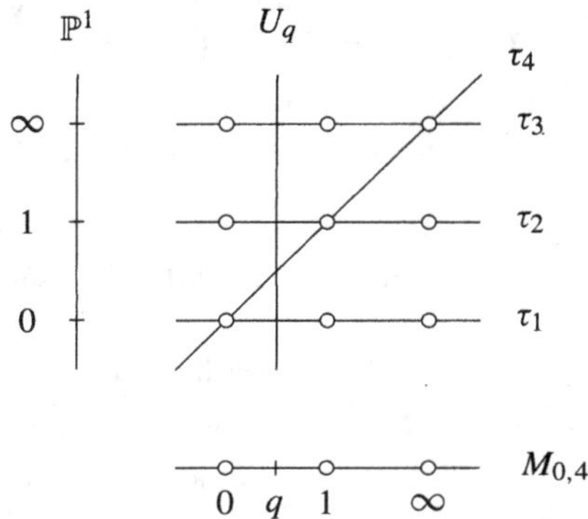

Note that $M_{0,4}$ is not compact. In 1.2.6 we will study its compactification $\overline{M}_{0,4}$.

1.1.5 Description of $M_{0,n}$ in general. To obtain $M_{0,5}$, take $M_{0,4} \times M_{0,4}$ and throw away the diagonal. In the same way it is not difficult to see that the space $M_{0,n}$ is isomorphic to the Cartesian product of $n-3$ copies of $\mathbb{P}^1 \smallsetminus \{0, 1, \infty\}$ *minus* all the big diagonals. (Think of it as the space parametrizing $n-3$ distinct cross ratios.)

$$M_{0,n} = \underbrace{M_{0,4} \times \cdots \times M_{0,4}}_{n-3} \smallsetminus \bigcup \text{diagonals}.$$

In particular, $M_{0,n}$ is smooth of dimension $n-3$. Note that the universal family is $U_{0,n} = M_{0,n} \times \mathbb{P}^1 \to M_{0,n}$, with the following sections. The first three are the constant ones $0, 1, \infty$ (in this order) and the remaining ones are induced by the $n-3$ projections $M_{0,n} \hookrightarrow M_{0,4} \times \cdots \times M_{0,4} \to M_{0,4} \subset \mathbb{P}^1$.

1.2 Stable n-pointed rational curves

A first idea to compactify $M_{0,n}$ is simply allowing the marked points to coincide; then the compactification would be something like $(\mathbb{P}^1)^{n-3}$ or $\mathbb{P}^{n-3}$. However, basic geometric properties would be lost with these compactifications, as the following example shows.

1.2.1 Example. Consider the two families of quadruples

$$C_t = (0, 1, \infty, t), \qquad D_t = (0, t^{-1}, \infty, 1).$$

For $t \neq 0, 1, \infty$, we have two families of 4-pointed smooth rational curves. Since the two families share the same cross ratio t, they are isomorphic. But the limits for $t = 0$ involve coincident points: C_0 has $p_1 = p_4$ (equal to zero), while D_0 has $p_2 = p_3$ (equal to infinity). Certainly these two configurations are not projectively equivalent.

The "correct" way to circumvent the anomaly just described was found by Knudsen and Mumford (cf. [51]). They showed that it is natural to include configurations where the curve "breaks." That is, we should admit certain reducible curves. The curves that appear in this good compactification are the *stable* curves we proceed to describe.

Definition. A *tree of projective lines*[1] is a connected curve with the following properties:

(i) Each irreducible component is isomorphic to a projective line.
(ii) The points of intersection of the components are ordinary double points.
(iii) There are no closed circuits. That is, if a node is removed, the curve becomes disconnected. Equivalently, if δ is the number of nodes, then there are $\delta + 1$ irreducible components.

The three properties together are equivalent to saying that the curve has arithmetic genus zero.

We will use the word *twig* for the irreducible components of such a tree, reserving the word *component* for the components of various subvarieties of the moduli space that will be considered throughout this book.

Definition. Let $n \geq 3$. A *stable* n-pointed rational curve is a tree C of projective lines, with n distinct marked points that are smooth points of C, such that every twig contains at least three special points. Here *special point* means a marked point or a node (point of intersection with another twig).

For short, we will often say *mark* instead of *marked point*, especially when we are more concerned with the name than with the actual point. All curves considered henceforth are rational. For this reason we will often simply say *stable n-pointed curve*, tacitly assuming that the curves are rational.

1.2.2 Example. In the figures below, all the twigs are isomorphic to $\mathbb{P}^1$. The first three curves are stable n-pointed rational curves, whereas the last four are not.

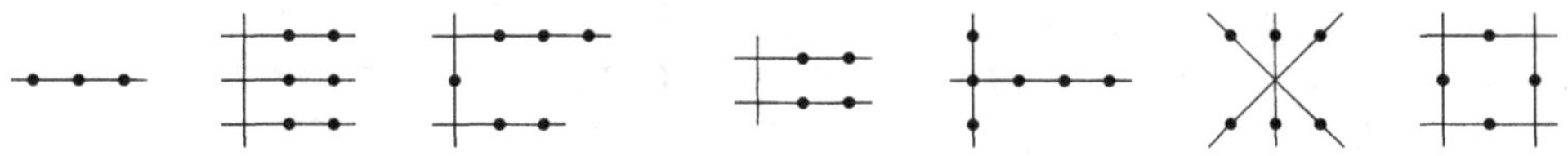

[1]It is somewhat abusive to call this a tree. It is really the dual graph that is a tree: associate a vertex to each irreducible component and an edge to each node. Then condition (ii) simply says that an edge is incident to precisely two vertices.

The fourth curve is not stable because the vertical twig has only two special points; the fifth curve is not stable because one of its marked points is a singular point of the curve. The sixth curve is not a tree in our sense of the word, since it has a triple point. Finally, the seventh curve is not a tree because it has a closed circuit. (However, it is stable as a 4-pointed curve of genus 1; cf. 1.6.5.)

1.2.3 Isomorphisms and automorphisms. An *isomorphism* of two n-pointed curves $(C, p_1, \ldots, p_n)$ and $(C', p'_1, \ldots, p'_n)$ is an isomorphism of curves $\phi : C \xrightarrow{\sim} C'$ such that $\phi(p_i) = p'_i$ for $i = 1, \ldots, n$. An *automorphism* of $(C, p_1, \ldots, p_n)$ is an automorphism $\phi : C \xrightarrow{\sim} C$ that fixes each marked point. We shall say that an n-pointed curve (or any given object or configuration) is *automorphism-free* if the identity is the only possible automorphism.

1.2.4 Stability in terms of automorphisms. If ϕ is an automorphism of a stable n-pointed curve $(C, p_1, \ldots, p_n)$ then since it fixes each marked point, in particular it maps each marked twig onto itself. Every twig with just one node must have marked points (by stability), and thus it is mapped onto itself, and since the node is the only singular point it must be a fixed point, and we conclude that the other twig is also mapped onto itself. By an induction argument, every node is fixed by ϕ and every twig is mapped onto itself. In other words, ϕ is obtained by gluing together automorphisms of each twig that fix all special points. But since there are at least three special points on each twig there can be no nontrivial automorphisms at all! Conversely, if there were a twig with less than three special points, there would be a nontrivial automorphism. So the stability condition is equivalent to saying that the n-pointed curve is automorphism-free.

A *family of stable n-pointed curves* is a flat and proper map $\pi : \mathfrak{X} \to B$ equipped with n disjoint sections, such that every geometric fiber $\mathfrak{X}_b := \pi^{-1}(b)$ is a stable n-pointed curve. In particular the sections are disjoint from the singular points of the fibers. The notion of isomorphism of two families with the same base is just as for smooth curves (cf. 1.1).

1.2.5 Theorem. (Knudsen [51]) *For each $n \geq 3$, there is a smooth projective variety $\overline{M}_{0,n}$ that is a fine moduli space for stable n-pointed rational curves. It contains the subvariety $M_{0,n}$ as a dense open subset.* □

In particular, the points of the variety $\overline{M}_{0,n}$ are in bijective correspondence with the set of isomorphism classes of stable n-pointed rational curves. The universal family $\overline{U}_{0,n} \to \overline{M}_{0,n}$ will be described below (1.4) as the morphism $\overline{M}_{0,n+1} \to \overline{M}_{0,n}$ defined by forgetting the last mark (cf. 1.3.5).

1.2.6 Example. Let us have a close look at the space $\overline{M}_{0,4}$ and its universal family $\overline{U}_{0,4} \to \overline{M}_{0,4}$. The only smooth compactification of $M_{0,4}$ is $\mathbb{P}^1$. However, if we simply put back the three missing points and correspondingly close up the total space of the universal family to get $\mathbb{P}^1 \times \mathbb{P}^1$ (see 1.1.4), we run into the problem that the sections will no longer be disjoint. Each of the three special fibers, e.g., the fiber U_0 over $q = 0$, has a doubly marked point, since τ_4 and τ_1 meet U_0 in one and the same point.

In order to remedy this situation, we blow up $\mathbb{P}^1 \times \mathbb{P}^1$ at these three bad points and set $\overline{U}_{0,4} := \mathrm{Bl}(\mathbb{P}^1 \times \mathbb{P}^1)$. Let E_0, E_1, and E_∞ be the exceptional divisors. This blowup fixes the problem: the fiber over $q = 0$ is now $\widetilde{U}_0 \cup E_0$, the union of two rational curves. See the figure:

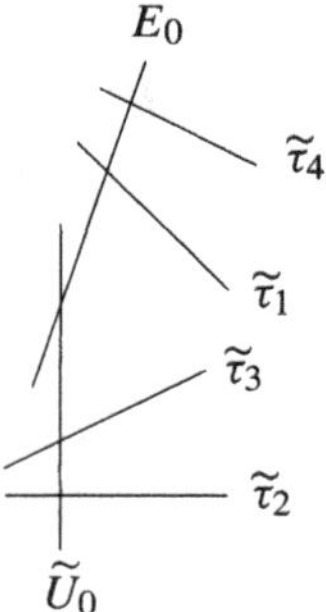

In the strict transform of the fiber, $\widetilde{U}_0$, there are three special points: the point of intersection with the exceptional divisor E_0, together with the two marked points given as the intersection with the strict transforms $\widetilde{\tau}_2$ and $\widetilde{\tau}_3$. These two marked points remain distinct since they are away from the blowup center; for the same reason, neither $\widetilde{\tau}_2$ nor $\widetilde{\tau}_3$ intersects E_0. In E_0 there are also two marked points, viz., the points of intersection with the strict transforms of τ_1 and τ_4. They are distinct since these two divisors intersect transversally in $\mathbb{P}^1 \times \mathbb{P}^1$.

In conclusion, the fiber over $q = 0$ consists of two projective lines meeting in an ordinary double point, and each of these lines carries two marked points. Hence there are three special points on each twig, and we have a stable curve.

The construction shows that whenever two marked points try to come together or coincide, a new twig sprouts out and receives the two marks. See the figure below.

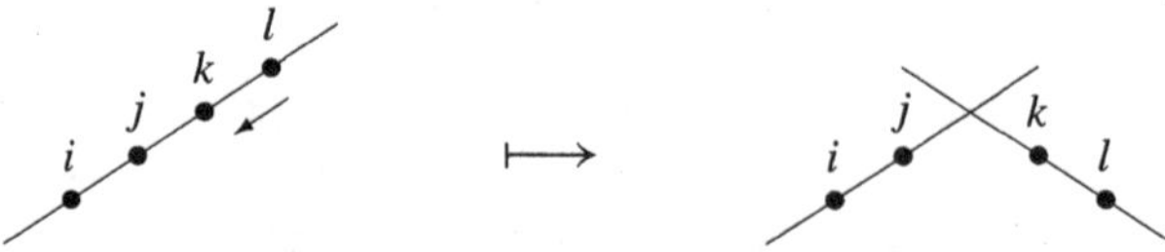

1.2.7 Example 1.2.1, revisited. Keeping the same notation, let us see how the problem of Example 1.2.1 is solved. The limit C_0 of the family $C_t = (0, 1, \infty, t)$ is the tree with two twigs such that $p_1 = 0$ together with p_4 are on one twig, and

$p_2 = 1$ and $p_3 = \infty$ on the other. Now note that up to (unique) isomorphism there is only one 4-pointed curve of this type. Indeed, there are exactly three special points on each twig, just as needed for the curve to be stable as well as for ruling out any freedom of choice. The same description goes for the limit D_0 of the family $D_t = (0, t^{-1}, \infty, 1)$. Consequently, these limits are equal, as desired.

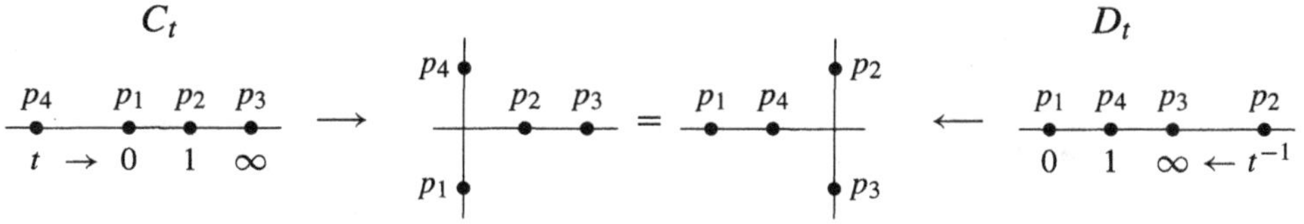

1.2.8 Remark. We saw that $\overline{U}_{0,4}$ is isomorphic to $\mathbb{P}^1 \times \mathbb{P}^1$ blown up at three points. Phrased differently, $\overline{U}_{0,4}$ is isomorphic to $\overline{M}_{0,4} \times_{\overline{M}_{0,3}} \overline{M}_{0,4}$ blown up at the three points. This viewpoint is the idea behind the generalization to the case of more marked points. See 1.4 below.

Note that in local analytic coordinates, in a neighborhood $\mathbb{A}^1 \simeq V \subset \overline{M}_{0,4}$ of a point $t = 0$ of the boundary of $\overline{M}_{0,4}$, the morphism $\overline{U}_{0,4} \to \overline{M}_{0,4}$ is of the form

$$\begin{array}{rcl} Q & \to & \mathbb{A}^1 \\ (x, y, t) & \mapsto & t, \end{array}$$

where $Q \subset \mathbb{A}^3$ is given by the equation $xy = t$. This observation also generalizes. See 1.4.4.

1.2.9 Remark. The blowup of $\mathbb{P}^1 \times \mathbb{P}^1$ at three points is isomorphic to the blowup of $\mathbb{P}^2$ at four points, which can be realized as the del Pezzo surface $S_5 \subset \mathbb{P}^5$. See Exercises 5–7 on page 43 for more details.

1.2.10 More marked points. In situations with more marked points, the figure for the degenerations is still practically the same. Whenever two marked points on a curve come together, a new (rational) twig sprouts out to receive the two marks. When many marked points are in play, it is also possible that three or more marked points come together simultaneously; in general this results in a single new rational twig on which the infringing marked points distribute themselves according to their ratio of approach to each other (as in (i) in the figure below). In special cases (if two such ratios coincide) the two points would be equal on the new twig, which of course is not allowed; instead the result is a whole new tree (as (ii) below). Finally, it can happen that one (or more) marked point approaches a node (the intersection of two twigs); then again a new twig arises to receive the infringing points, as pictured in (iii):

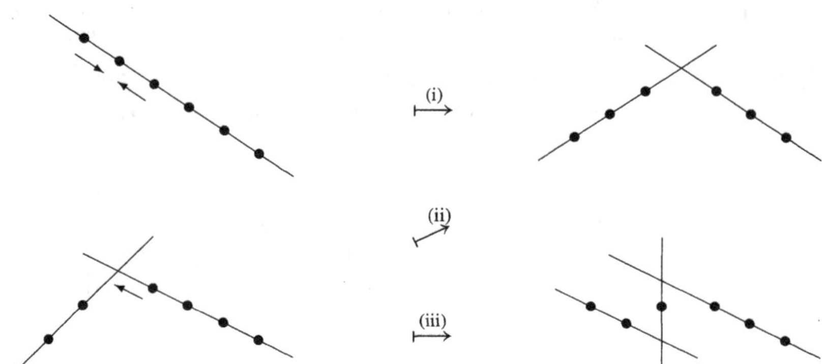

1.2.11 Remark. One could ask what happens when two marked points collide on a twig with only three special points. If this were possible and a new twig sprouted out, it would follow that only two special points would be left on the original twig and the stability would be lost. But note that this behavior cannot happen: when one of the three special points begins to move, the curve stays isomorphic to the original curve, since there exists an automorphism that sends the new configuration back to the original one. This way, moving around the marks on a twig with just three special points does not draw a curve in the moduli space: we never leave the same original moduli point.

1.3 Stabilization, forgetting marks, contraction

1.3.1 Stabilization. Given a stable n-pointed curve $(C, p_1, \ldots, p_n)$ and an arbitrary point $q \in C$, we are going to describe a canonical way to produce a stable $(n+1)$-pointed curve. If q is not a special point, then we just set $p_{n+1} := q$ and we have an $(n+1)$-pointed curve that is obviously stable. If q is a special point of C, we may initially set $p_{n+1} := q$; but then this $(n+1)$-pointed curve is not stable. What we claim is that there is a canonical way of *stabilizing* a pointed curve of this type. We have already seen examples in the previous section hinting at how this must be done: to preserve continuity in the stabilization process, the limit of the stabilization must be the stabilization of the limit. If $p_{n+1} = q$ runs through nonspecial points of C and suddenly coincides with a node or a marked point p_i, then we have already seen what the limit is, and thus, what the stabilization should be. Let us spell out the two cases:

Case I. If q is the node of intersection of two twigs, pull them apart and form a new curve putting in a new twig joining the two points and put the new mark p_{n+1} on this new twig.

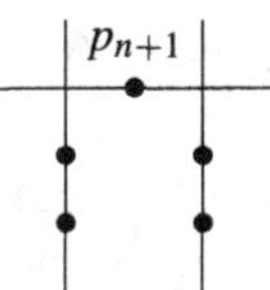

is the stabilization of
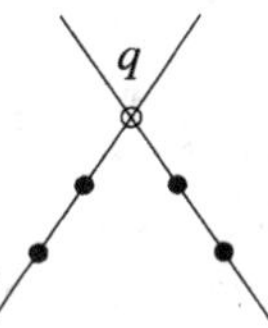

Case II. If q coincides with one of the marked points, say p_i, create a new twig at this point and put both marks p_i and p_{n+1} on it.

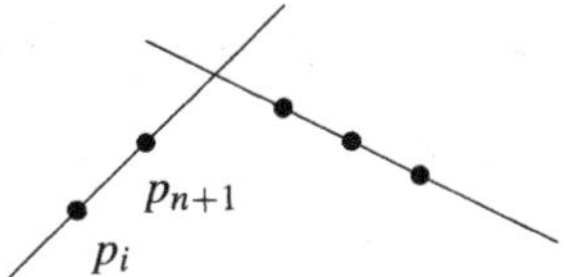

is the stabilization of
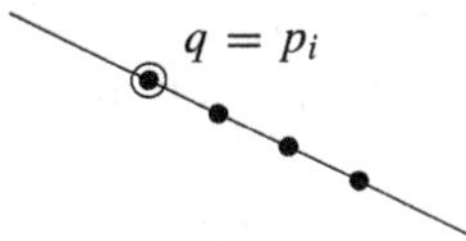

Note that in both cases, the choice of the position of the special points on the new twig is irrelevant, since there are exactly three special points on this twig.

More precisely, this process also works in families:

1.3.2 Proposition. (Knudsen [51]) *Given a family of stable n-pointed curves* $(\mathfrak{X}/B, \sigma_1, \dots \sigma_n)$*, let* $\delta : B \to \mathfrak{X}$ *be an arbitrary extra section. Then there exists a family* $(\mathfrak{X}'/B, \sigma_1', \dots, \sigma_n', \sigma_{n+1}')$ *of stable* $(n+1)$*-pointed curves (over B) and a B-morphism* $\varphi : \mathfrak{X}' \to \mathfrak{X}$ *such that*

(i) its restriction $\varphi^{-1}(\mathfrak{X} \setminus \delta) \xrightarrow{\sim} \mathfrak{X} \setminus \delta$ *is an isomorphism,*

(ii) $\varphi \circ \sigma_{n+1}' = \delta$,

(iii) $\varphi \circ \sigma_i' = \sigma_i$*, for* $i = 1, \dots, n$.

Up to isomorphism, this family is unique with these properties; it is called the stabilization *of* $(\mathfrak{X}/B, \sigma_1, \dots \sigma_n, \delta)$.

Furthermore, this stabilization commutes with fiber products. □

The last assertion is useful because it ensures that the fibers of the stabilization are the stabilizations of the fibers. In particular, for all $b \in B$ such that $\delta(b)$ is distinct from each $\sigma_i(b)$, the fibers $\mathfrak{X}_b'$ and $\mathfrak{X}_b$ are isomorphic as stable $(n+1)$-pointed curves.

We already saw an example of stabilization in a family, namely in 1.2.6, where the extra section was the diagonal. The stabilization consisted in blowing up the three points of intersection among the sections. This observation generalizes, and Knudsen [51] shows how the stabilization is given by specific blowups.

1.3.3 Forgetting marks and contraction. Conversely, given a stable $(n+1)$-pointed curve $(C, p_1, \dots, p_n, p_{n+1})$ there exists a canonical way of associating to it a stable n-pointed curve (assuming $n \geq 3$). The first step is simply removing p_{n+1}. This yields an n-pointed curve $(C, p_1, \dots, p_n)$. If C is an irreducible curve, then obviously the resulting n-pointed curve is stable. But in the case where C

is reducible, removing p_{n+1} can destabilize a twig and render $(C, p_1, \dots, p_n)$ unstable. What we claim is that there is a canonical way of stabilizing this curve. Since we already used the term stabilization, we will call this process *contraction* (following the terminology of Knudsen [51]). What does the job now is simply contracting any twig that has become unstable.

We will denote by the term *forgetting* p_{n+1} the two-step process of removing p_{n+1} and then contracting any unstable twig if such appears. Let us draw some figures to illustrate what happens when the mark p_{n+1} is erased.

Case I. If p_{n+1} is on a twig without other marked points, and with just two nodes, then this twig is contracted:

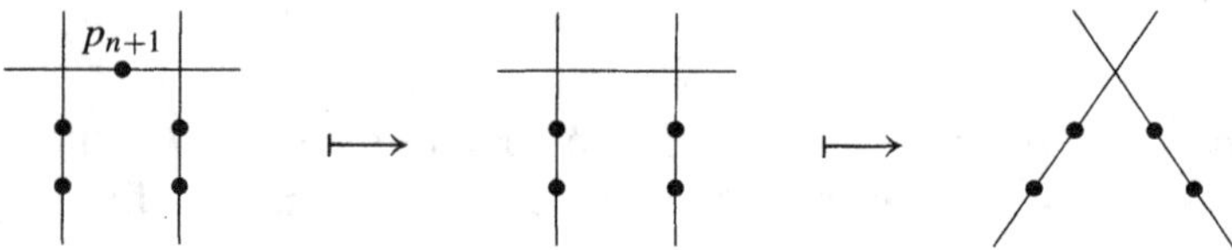

Case II. If p_{n+1} is on a twig with just one other marked point p_i, and only one singular point (the point where the twig is attached to the rest of the curve), then the twig is contracted and the point where the twig was attached acquires the mark p_i:

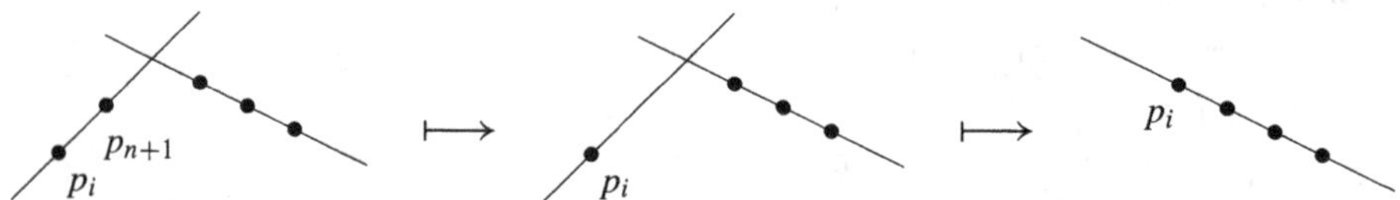

Except for these two cases, forgetting marks does not involve contraction.

As in the case of stabilization, everything works fine in families:

1.3.4 Proposition. (Knudsen [51]) *Let $(\mathfrak{X}'/B, \sigma'_1, \dots, \sigma'_n, \sigma'_{n+1})$ be a family of stable $(n+1)$-pointed curves. Then there exists a family $(\mathfrak{X}/B, \sigma_1, \dots, \sigma_n)$ of stable n-pointed curves equipped with a B-morphism $\varphi : \mathfrak{X}' \to \mathfrak{X}$ such that*

(i) $\varphi \circ \sigma'_i = \sigma_i$, for $i = 1, \dots, n$;

(ii) for each $b \in B$, the induced morphism $\mathfrak{X}'_b \to \mathfrak{X}_b$ is an isomorphism when restricted to any stable twig of $(\mathfrak{X}'_b, \sigma'_1(b), \dots, \sigma'_n(b))$, and it contracts an eventual unstable twig.

The family $(\mathfrak{X}/B, \sigma_1, \dots, \sigma_n)$ is unique up to isomorphism, and we shall say that it is the family obtained from $\mathfrak{X}'/B$ by forgetting σ'_{n+1}.

Furthermore, forgetting sections commutes with fiber products. □

1.3.5 Remark. We have described the map $\overline{M}_{0,n+1} \to \overline{M}_{0,n}$ set-theoretically. The proposition guarantees that it is in fact a morphism. Indeed, consider the universal family of stable $(n+1)$-pointed curves $\overline{U}_{0,n+1} \to \overline{M}_{0,n+1}$. Forgetting the last mark yields a family of stable n-pointed curves, with the same base $\overline{M}_{0,n+1}$. Now the universal property of $\overline{M}_{0,n}$ gives us a morphism

$$\varepsilon : \overline{M}_{0,n+1} \to \overline{M}_{0,n},$$

which clearly coincides with the set-theoretical description we already had. This morphism is called the *forgetful map*.

1.3.6 Remark. In the cases described above we forgot the last mark. However, this choice was only for notational convenience. We could equally well forget any other mark: all the marks are on equal footing. See 1.5.12 below, where we combine various forgetful maps.

1.3.7 Comparison between stabilization and contraction. It is worth noticing that the curve obtained by forgetting the mark p_{n+1} always comes with a distinguished point (not a marked point), namely the point $\varphi(p_{n+1})$. In the general case, where no contraction is involved, this distinguished point is a nonspecial point. In the two cases where contraction occurs, the distinguished point is a special point, either a node (case I), or a marked point (p_i as in case II). See the diagram (for case I):

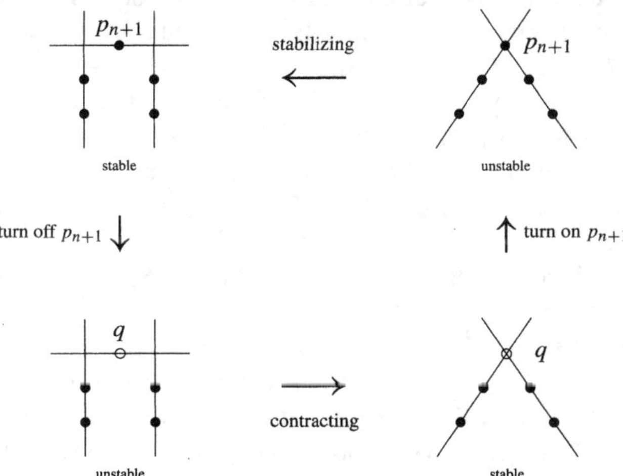

Here the point q is not considered a marked point. The direction of the arrows indicates only the "direction of the construction," and is not meant to say that there exists a morphism. (The direction of the morphism is always rightward and downward.)

We could also draw a diagram for case II (the situation in which q is on top of a marked point p_i). We leave that task to the reader.

1.4 Sketch of the construction of $\overline{M}_{0,n}$

The key observation for the construction of $\overline{M}_{0,n}$ as a fine moduli space is that there is an isomorphism $\overline{M}_{0,n+1} \simeq \overline{U}_{0,n}$. In this way, the construction is iterative. Follow the diagram below, starting from the bottom:

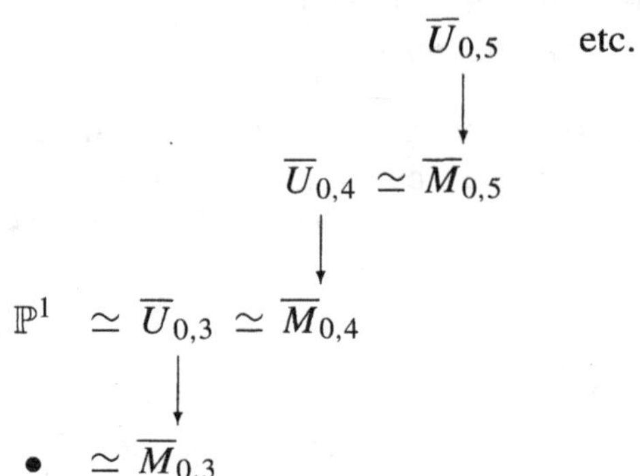

Let us argue why $\overline{U}_{0,4} \simeq \overline{M}_{0,5}$, and next construct the universal family $\overline{U}_{0,5} \to \overline{M}_{0,5}$. The procedure is similar for $n \geq 5$.

1.4.1 (Set-theoretical) construction of $\overline{M}_{0,5}$. The first step is to establish a natural bijection of sets between $\overline{U}_{0,4}$ and $\overline{M}_{0,5}$. To each point $q \in \overline{U}_{0,4}$ we shall associate a stable 5-pointed curve C_q. Let us denote by $\pi : \overline{U}_{0,4} \to \overline{M}_{0,4}$ the universal family. Given $q \in \overline{U}_{0,4}$, write $F_q = \pi^{-1}\pi(q)$ for the fiber passing through q. In other words, $\pi(q) \in \overline{M}_{0,4}$ represents a stable 4-pointed curve isomorphic to the fiber F_q. Now the point q itself singles out a fifth marked point yielding in this way a 5-pointed curve that we denote by (F_q, q). In case q is not a special point of F_q, this curve is automatically stable and we can call it C_q, the promised stable 5-pointed curve. If the point q is a special point of F_q, then we take as C_q the *stabilization* of (F_q, q).

It is clear that the map $\overline{U}_{0,4} \ni q \mapsto C_q \in \overline{M}_{0,5}$ is injective. On the other hand, given any stable 5-pointed curve $(C, p_1, \dots, p_5)$, we can forget p_5 to get a stable 4-pointed curve, together with a point on it (the place where p_5 was; cf. 1.3.7). This specifies a fiber of π together with a point in the fiber, i.e., a point in $\overline{U}_{0,4}$. This is the asserted bijection. Note that by construction, the morphism π is identified with the morphism forgetting p_5 (cf. 1.3.5). Under the set-theoretical bijection, the fiber of the forgetful map is identified with the fiber of π.

Let us now sketch the second ingredient in the iterative procedure: the construction of the universal curve $\overline{U}_{0,5}$ over $\overline{M}_{0,5}$. This means that $\overline{M}_{0,5}$ is not merely in set-theoretical bijection with the set of isomorphism classes of stable curves; it is indeed a fine moduli space.

1.4.2 Construction of the universal family $\overline{U}_{0,5}$. As in the case considered above, $\pi : \overline{U}_{0,4} \to \overline{M}_{0,4}$, the idea is to take the fiber product over $\overline{M}_{0,4}$ of two copies of $\overline{M}_{0,5}$ and then stabilize.

Consider the Cartesian diagram

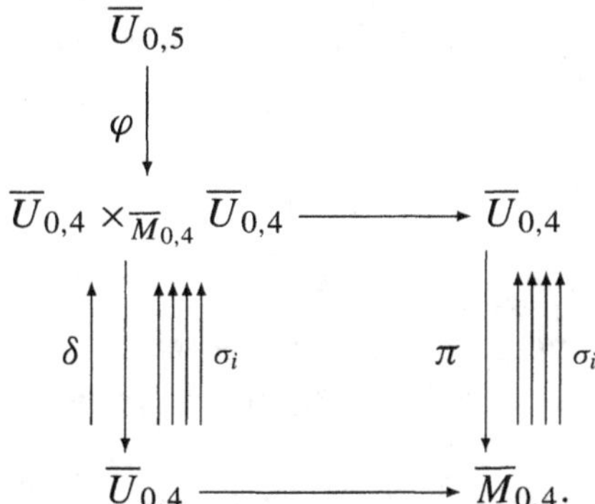

The pullback of π, together with the maps induced by the sections σ_i, constitute a family of 4-pointed curves parametrized by $\overline{U}_{0,4}$ (left-hand side of the diagram). This family admits yet another natural section, namely the diagonal section δ. This section destabilizes the family since it is not disjoint from the other four sections. But we know from 1.3.1 that there is a stabilization. We claim that the stabilization is a universal family and therefore we will denote it by $\overline{U}_{0,5}$.

Let us show that the fiber $(\overline{U}_{0,5})_q$ of this family over a point $q \in \overline{U}_{0,4}$ is a 5-pointed curve isomorphic to C_q (cf. 1.4.1). The fiber over q is the pullback of the fiber over $\pi(q) \in \overline{M}_{0,4}$. But we know that the fiber over a point in $\overline{M}_{0,4}$ is isomorphic to the curve it represents, in the case at hand the 4-pointed curve F_q. Now by construction, there is yet another marked point on this curve, given by the diagonal section δ. That is, the fifth marked point is $q \in F_q$ itself. Therefore, the fiber over q of $\overline{U}_{0,4} \times_{\overline{M}_{0,4}} \overline{U}_{0,4} \to \overline{U}_{0,4}$ is the 5-pointed curve (F_q, q), which is not necessarily stable. What we wanted was the fiber of the stabilization $\overline{U}_{0,5}$. But we know that stabilization commutes with fiber products: the fiber of the stabilization is the stabilization of the fiber. By the very construction of the set-theoretical bijection, the stabilization of (F_q, q) is exactly the stable 5-pointed curve C_q.

This shows that $\overline{M}_{0,5}$ (equipped with the scheme structure from $\overline{U}_{0,4}$) possesses a tautological family. To establish that it is a fine moduli space it remains to show that the family enjoys the universal property. We omit this verification.

1.4.3 Remark. We should stress that the stabilization of the family $\overline{U}_{0,4} \times_{\overline{M}_{0,4}} \overline{U}_{0,4} \to \overline{U}_{0,4}$ is not simply a blowup along the intersections of the sections as it was the case for $n = 3$. For $n \geq 4$, it is subtler, since the morphism $U_{0,4} \times_{\overline{M}_{0,4}} \overline{U}_{0,4} \to \overline{U}_{0,4}$ is not smooth. Thus δ is no longer a regular embedding, and the blowup becomes a singular variety. But it is possible to blow up a bit more, and the minimal desingularization will be the sought-for stabilization $\overline{U}_{0,5}$.

1.4.4 Local description of π. In $\mathbb{A}^3$, with coordinates (x, y, t), consider the quadric Q given by the equation $xy = t$. We already mentioned that the morphism $\overline{U}_{0,4} \to \overline{M}_{0,4}$, local-analytically around a point of the boundary of $\overline{M}_{0,4}$, is of the form

$$\begin{array}{rcl} Q & \to & \mathbb{A}^1 \\ (x, y, t) & \mapsto & t. \end{array}$$

In particular, it has reduced geometric fibers.

In the same way, Knudsen shows that the morphism $\overline{U}_{0,n} \to \overline{M}_{0,n}$ local-analytically around a point of the boundary of $\overline{M}_{0,n}$ is of the form

$$\begin{array}{rcl} V \times Q & \to & V \times \mathbb{A}^1 \\ \big(v, (x, y, t)\big) & \mapsto & (v, t) \end{array}$$

where V is a smooth variety. Its geometric fibers are reduced.

1.5 The boundary

Each point in the boundary $\overline{M}_{0,n} \smallsetminus M_{0,n}$ corresponds to a reducible curve.

1.5.1 The stratification of $\overline{M}_{0,n}$. We first notice that the subset Σ_δ of $\overline{M}_{0,n}$ consisting of curves with $\delta \leq n - 3$ nodes is of pure dimension $n - 3 - \delta$.

The argument is a simple dimension count. We can compute the dimension by summing the degrees of freedom of each twig (freedom of moving marked points and nodes). Together we have $n + 2\delta$ special points, since each node is a special point on each of the two twigs that intersect in it. Now by stability, each twig has at least three special points, and we know there exists an automorphism that sends these three points to $0, 1, \infty$. That is, on each twig, three of the special points are spent with getting rid of automorphisms. Since C is a tree, the number of twigs is $\delta + 1$. So we conclude that

$$\dim \Sigma_\delta = n + 2\delta - 3(\delta + 1) = n - 3 - \delta$$

as claimed.

The justification for counting parameters twig by twig will be given in 1.5.9, where we will see that the stratum in $\overline{M}_{0,n}$ corresponding to curves with δ nodes is locally a product of the moduli spaces of its twigs.

1.5.2 Example. Here is a drawing of the stratification of $\overline{M}_{0,6}$. The six marked points have not been assigned names, but the number next to each figure indicates how many ways there are to label the given configuration:

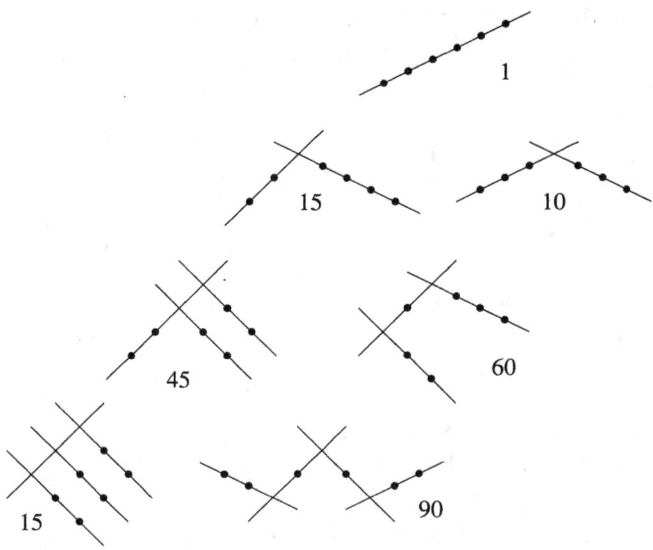

Each number appears as a multinomial coefficient, divided by the number of symmetries of the configuration. For example, the last number is $90 = \left(\frac{6!}{2!\,1!\,1!\,2!}\right)/2$ because the configuration is symmetric in the middle.

So there is a single stratum of maximal dimension 3 (the dense subvariety $M_{0,6} \subset \overline{M}_{0,6}$); 25 strata in dimension 2; 105 strata in dimension 1; and by coincidence also 105 strata in dimension 0.

1.5.3 Boundary cycles. The closure of each given labeled configuration is a (smooth and irreducible) subvariety of $\overline{M}_{0,n}$, called a *boundary cycle*. As suggested by the previous figure, the boundary of a boundary cycle is made up of boundary cycles (of higher codimension) corresponding to configurations that are further degenerated.

Irreducibility and smoothness of each boundary cycle will be established in 1.5.9 below, when we study the recursive structure of the boundary.

1.5.4 Boundary divisors. Particularly interesting are the *boundary divisors*. They are the boundary cycles of codimension 1.

Denote by S the marking set $\{p_1, p_2, \ldots, p_n\}$. There is an irreducible boundary divisor $D(A|B)$ for each partition $S = A \cup B$ with A, B disjoint and $\sharp A \geq 2$, $\sharp B \geq 2$. A general point of $D(A|B)$ represents a curve with two twigs, with the marks of A on one twig, and the marks of B on the other.

At the boundary of each boundary divisor we find the possible degenerations of the given configuration, that is, boundary cycles of higher codimension.

1.5.5 Example. Consider the intersection of $D(ab|cdef)$ with $D(abc|def)$ inside $\overline{M}_{0,6}$. We see that the only degenerations common to the two divisors are those contained in the (closure of) the configuration $(ab|c|def)$ as indicated in the following figure:

The moral of this example is that the intersection of two irreducible boundary divisors is always an irreducible codimension-2 stratum, except when it is empty, as we see in the remark below.

1.5.6 Remark. Let $A \cup B = S$ and $A' \cup B' = S$ be two partitions of the marking set S such that there are no inclusions among any two of the four sets A, B, A', B'. In this case,

$$D(A|B) \cap D(A'|B') = \emptyset.$$

Indeed, there are no common degenerations.

1.5.7 Example. By simple combinatorics (cf. Exercise 9 on page 45) there are 10 boundary divisors in $\overline{M}_{0,5}$. We also find 10 divisors in $\overline{U}_{0,4}$ (4 sections, 3 exceptional divisors, and the 3 strict transforms of the special fibers). Let us see how these divisors match up under the identification $\overline{U}_{0,4} \simeq \overline{M}_{0,5}$. A point $q \in \overline{U}_{0,4}$ (see 1.2.6) on a section σ_i (say σ_1) parametrizes a curve whose fifth marked point q "coincides" with the first marked point, or more precisely, these two marked points are together on one and the same twig. Therefore the divisor $\sigma_1 \subset \overline{U}_{0,4}$ corresponds to the boundary divisor of $\overline{M}_{0,5}$ whose general curve is of type indicated in the figure:

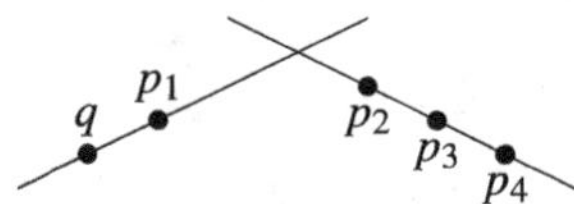

A point $q \in E_0$ maps to the point 0 of the boundary of $\overline{M}_{0,4}$. Let us say that $0 \in \overline{M}_{0,4}$ corresponds to the partition $(p_1, p_2 \mid p_3, p_4)$. It follows that the divisor $E_0 \subset \overline{U}_{0,4}$ corresponds to the boundary divisor of $\overline{M}_{0,5}$ whose description is:

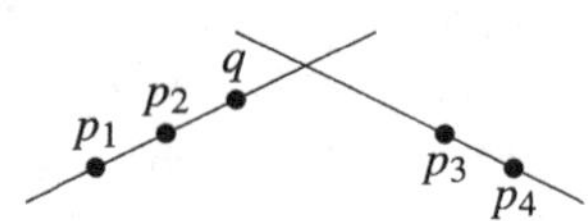

The same argument shows that the strict transform of the fiber F_0 is the boundary divisor whose general member is of the form

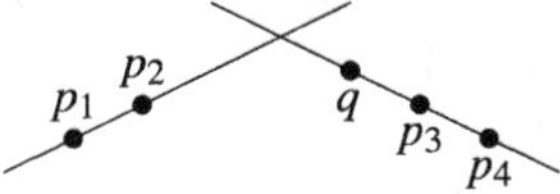

Finally, let us describe the point $q \in E_0 \cap F_0$. It corresponds to the intersection of the two divisors, that is, to the following configuration:

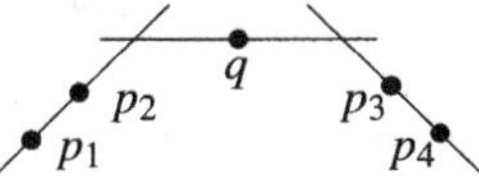

It is clear from the blowup description of $\overline{M}_{0,5}$ that the 10 boundary divisors intersect transversally. This behavior proliferates up through the hierarchy of $\overline{M}_{0,n}$'s, as the following proposition states.

1.5.8 Proposition. *The boundary of $\overline{M}_{0,n}$ is a divisor with normal crossings.* □

1.5.9 The recursive structure. Each boundary cycle is naturally isomorphic to a product of moduli spaces of lower dimension. Let us study in detail the case of a boundary divisor $D(A|B)$.

A general point of $D(A|B)$ corresponds to a reducible curve R with two twigs, with the marks of A distributed on one twig and the marks of B on the other. Now take each twig separately and denote the point of intersection with the other twig by the letter x. The A-twig gives us an element of $\overline{M}_{0,A\cup\{x\}}$, and the B-twig gives an element in $\overline{M}_{0,B\cup\{x\}}$. Note that stability of R implies (indeed, is equivalent to) stability of each of these two curves.

Conversely, every element in $\overline{M}_{0,A\cup\{x\}} \times \overline{M}_{0,B\cup\{x\}}$ reproduces a curve of the original configuration, identifying the two points marked x, attaching the two curves in a node at this point. We get in this way a canonical isomorphism

$$D(A|B) \simeq \overline{M}_{0,A\cup\{x\}} \times \overline{M}_{0,B\cup\{x\}},$$

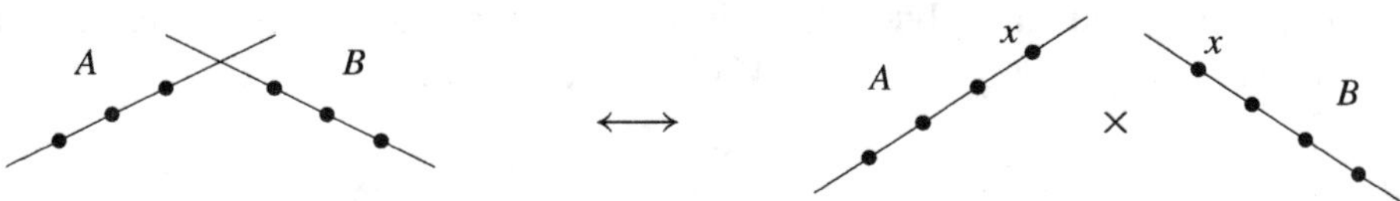

In particular, knowing that the moduli spaces with fewer marks are irreducible and smooth, we conclude that the boundary divisors $D(A|B)$ are irreducible and smooth as well. Similar arguments apply to any boundary cycle: the cycle corresponding to a labeled configuration (say with δ nodes) is naturally isomorphic to a product of $\delta + 1$ moduli spaces, and therefore in particular irreducible and smooth. It also follows from this description that it is legal to perform the "twig-by-twig" dimension count given in 1.5.1.

1.5.10 Pullback of boundary divisors under forgetful maps. Consider the forgetful map $\varepsilon : \overline{M}_{0,n+1} \to \overline{M}_{0,n}$, which forgets the last mark q. Let $D(A|B)$ be a boundary divisor of $\overline{M}_{0,n}$. Then in the inverse image there are two possibilities: either the extra mark q is on the A-marked twig, or it is on the B-marked. This describes the inverse image set-theoretically. Recalling now that the geometric fibers are reduced (cf. 1.4.4), we conclude that the coefficients of the pullback divisor are equal to 1, i.e.,

$$\varepsilon^* D(A|B) = D(A \cup \{q\} \mid B) + D(A \mid B \cup \{q\}).$$

It should also be noted that the recursive structure is compatible with forgetful maps. This means that diagrams like this one commute:

$$\begin{array}{ccccc}
\overline{M}_{0,A\cup\{x\}} \times \overline{M}_{0,B\cup\{q,x\}} & \stackrel{\sim}{\to} & D(A|B \cup \{q\}) & \subset & \overline{M}_{0,n+1} \\
\varepsilon \downarrow & & \downarrow \varepsilon_{|D} & & \downarrow \varepsilon \\
M_{0,A\cup\{x\}} \times \overline{M}_{0,B\cup\{x\}} & \stackrel{\sim}{\to} & D(A|B) & \subset & \overline{M}_{0,n}
\end{array}$$

where each of the vertical arrows is forgetting q, the last mark.

1.5.11 Example. The n sections of the universal curve $\overline{U}_{0,n} \to \overline{M}_{0,n}$ admit an interpretation in terms of the forgetful map $\varepsilon : \overline{M}_{0,n+1} \to \overline{M}_{0,n}$. Recall that the ith section is the one that "repeats the ith marked point" and stabilizes. So the image of section σ_i is the boundary divisor

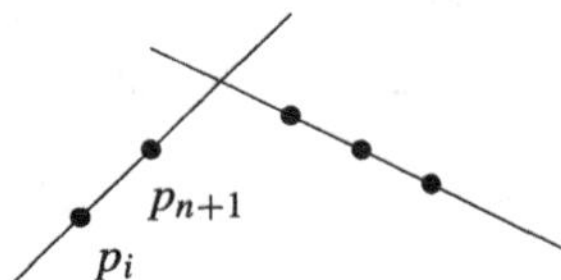

(We are talking about case II described in 1.3.1 on page 28.) With this observation, we can compare the boundary F_n of $\overline{M}_{0,n}$ with the boundary F_{n+1} of $\overline{M}_{0,n+1}$:

$$F_{n+1} = \varepsilon^* F_n + \sum \sigma_i,$$

where by abuse of notation, the symbol σ_i also denotes the image in $\overline{M}_{0,n+1}$ of that section.

1.5.12 Composing forgetful maps. As remarked in 1.3.6, nothing prevents us from forgetting marked points other than the last one. Suppose that $S = \{p_1, \ldots, p_n\}$ is the set of marks, and let $B \subset S$, $\#B \geq 3$. Then there is a morphism $\overline{M}_{0,S} \to \overline{M}_{0,B}$ given by forgetting all the marks in the complement $S \smallsetminus B$. It is simply the composition of the forgetful maps studied in 1.3.5. Note that all these morphisms commute, in the sense that it does not matter in which order we forget the marked points. This is clear when restricted to the dense open set of smooth curves, and the general statement follows from the compatibility of the forgetful maps with the recursive structure described in 1.5.10.

1.5.13 Special boundary divisors. Particularly important is the forgetful map $\overline{M}_{0,n} \to \overline{M}_{0,4} = \mathbb{P}^1$, assuming $n \geq 4$. Pick one of the three boundary divisors of $\overline{M}_{0,4}$, say $D(ij|kl)$: its pullback to $\overline{M}_{0,n}$ is a sum of boundary divisors $D(A|B)$. Combining the formulas 1.5.10 for each step in the composition of forgetful maps, we see that the result is the sum over all partitions $A \cup B = S$ such that $i, j \in A$ and $k, l \in B$. The formulas also guarantee that all the coefficients in the sum are equal to 1.

From the obvious fact that any two points on $\overline{M}_{0,\{i,j,k,l\}} \simeq \mathbb{P}^1$ are linearly equivalent, it follows that their pullbacks in $\overline{M}_{0,n}$ are also linearly equivalent:

$$\sum_{\substack{i,j \in A \\ k,l \in B}} D(A|B) \equiv \sum_{\substack{i,k \in A \\ j,l \in B}} D(A|B) \equiv \sum_{\substack{i,l \in A \\ j,k \in B}} D(A|B). \tag{1.5.13.1}$$

These relations will play a decisive role in this book.

1.5.14 The Chow ring of $\overline{M}_{0,n}$. Keel [47] shows that the classes of the boundary divisors $D(A|B)$ generate the Chow ring, and that a complete set of relations is provided by the relations described in the preceding paragraph together with those of 1.5.6.

1.6 Generalizations and references

1.6.1 Kapranov's construction. The moduli space $\overline{M}_{0,n}$ has another geometric interpretation (due to Kapranov [46]) which is worth mentioning. It is well known that through $n + 1$ general points in $\mathbb{P}^{n-2}$ there is a unique rational normal curve (cf. Exercise 3 on page 43), e.g., a unique smooth conic through 5 points in $\mathbb{P}^2$ (cf. also 3.2.2), and a unique twisted cubic through 6 points in $\mathbb{P}^3$. Now instead of $n + 1$ general points we fix only n points. Note that any independent set of n points in $\mathbb{P}^{n-2}$ is projectively equivalent to the coordinate points; cf. Exercise 3 on page 16. Then there is an $(n - 3)$-dimensional family of all rational normal curves through these n points. There are also reducible degree-$(n - 2)$ curves (trees of rational curves) that go through all the points, but no nonreduced curves. Kapranov shows that the family of all these curves is naturally isomorphic to $\overline{M}_{0,n}$. The basic observation is that a $\mathbb{P}^1$ with n marked points $P_1, \ldots, P_n$ is embedded in $\mathbb{P}^{n-2}$ by the linear system $\left|K_{\mathbb{P}^1} + P_1 + \cdots + P_n\right| = |\mathcal{O}(n-2)|$. Now the rough idea is that similarly the universal curve $\overline{U}_{0,n}$ embeds into a $\mathbb{P}^{n-2}$-bundle over $\overline{M}_{0,n}$; this bundle can be trivialized since it has n disjoint sections (which turn out always to be general points of the fiber), so altogether we get a morphism $\overline{U}_{0,n} \to \mathbb{P}^{n-2}$ that sends each section to one of the specified points. In this way every moduli point in $\overline{M}_{0,n}$ defines a degree-n curve through the n points in $\mathbb{P}^{n-2}$, and conversely every such curve has the n points as marks and defines a moduli point. Using the isomorphism $\overline{M}_{0,n+1} \simeq \overline{U}_{0,n}$ we get a morphism $\overline{M}_{0,n+1} \to \mathbb{P}^{n-2}$ that is birational. In fact it is a specific blowup along coordinate points, coordinate lines, etc., and hence we get an alternative explicit description of the moduli space.

The $n = 5$ case of Kapranov's construction is treated in the exercises.

The construction and results of this chapter have analogues for curves of positive genus, but the theory is much subtler. The case of rational curves is very special, in that any two rational curves are isomorphic; thus the theory of moduli is mostly concerned with the configuration of marked points.

1.6.2 Moduli of curves. It was known to Riemann [71] that the isomorphism classes of smooth curves of genus $g \geq 2$ constitute a family of dimension $3g - 3$ (in Riemann's words, the collection depends on $3g - 3$ complex modules; this is the origin of the term moduli space). The starting point for the modern approach to moduli of curves is Grothendieck's work on Hilbert schemes [41], and Mumford's geometric invariant theory [62]. Mumford showed that for each $g \geq 2$ there is a coarse moduli space M_g of dimension $3g-3$ parametrizing isomorphism classes of smooth curves of genus g, and (joint with Mayer) identified which curves are needed in order to compactify the moduli space: they are the *stable curves*: connected curves with ordinary double points as worst singularities. The work of Mayer and Mumford remained unpublished (a historical account can be found in Appendix 5D of the second or third edition of G.I.T. [62]). The general construction of the compactifcation $\overline{M}_g$ was given a few years later by Deligne and Mumford [16]. They showed in particular that $\overline{M}_g$ is a coarse moduli space for isomorphism classes of stable curves. $\overline{M}_g$ is smooth off the locus of curves with automorphisms, and locally it is a quotient of a smooth variety by a finite group.

A good starting point is the recent book by Harris and Morrison [43].

1.6.3 Elliptic curves. Smooth curves of genus 1 (elliptic curves) are classified by the j-invariant (cf. Hartshorne [44, Ch. 4]; see also Exercise 8 on page 18), but here we are really talking about 1-pointed curves, the marked point being the origin of the elliptic curve. So $M_{1,1}$ is isomorphic to $\mathbb{A}^1$, and in the compactification $\overline{M}_{1,1} \simeq \mathbb{P}^1$, the point at infinity corresponds to the nodal rational curve (of arithmetic genus 1).

1.6.4 Intersection theory on M_g and $\overline{M}_g$ was first undertaken by Mumford [64]. There are certain *tautological classes*, which are of particular interest: they are defined in terms of the relative dualizing line bundle ω_π on the universal curve $\pi : C_g \to M_g$. Since the universal curve does not exist over the locus of curves with automorphisms, Mumford used intersection theory on the *moduli functor*, this is a trick that roughly amounts to doing intersection theory on every family at once. Taking the various powers of the first Chern class of ω_π, and then pushing these classes down in M_g defines the *kappa classes*. Taking instead the direct image sheaf $\pi_* \omega_\pi$ and taking Chern classes of this rank-g vector bundle defines the *lambda classes*. Mumford used Grothendieck–Riemann–Roch to establish relations between these classes, and showed how to express the class of many geometric loci (e.g., the locus of hyperelliptic curves) in terms of kappa and lambda classes (and the boundary classes if we are on $\overline{M}_g$).

In addition to [43], let us also recommend the notes of Gatto [33], where many detailed examples and calculations of this sort are found.

1.6.5 Deligne–Mumford–Knudsen spaces. The pointed version of the Deligne–Mumford moduli spaces were introduced by Knudsen [51] in the late seventies, and they are usually called Deligne–Mumford–Knudsen spaces. However, Grothendieck had already studied such spaces already in 1968 (cf. [15]), and sometimes they are also called Grothendieck–Knudsen spaces, or Grothendieck–Mumford–Knudsen spaces. Knudsen introduced marked points as an auxiliary structure, used to prove the projectivity of the unpointed spaces $\overline{M}_g$. Later it has turned out, however, that the pointed spaces $\overline{M}_{g,n}$ are also very important in their own right; cf. below. Here are the main results of Knudsen — the original paper [51] remains the main reference.

A *stable n-pointed curve* is a connected n-pointed curve C (as usual the marked points must be distinct and smooth), with ordinary double points as worst singularities, subject to the following stability condition: every genus-0 component must have at least three special points and every genus-1 component must have at least one special point; this last condition only serves to rule out genus-1 curves without marked points. Note that stable n-pointed curves of genus $g > 0$ can have nontrivial automorphisms; the stability condition is equivalent to requiring the curve to have only finitely many automorphisms. There exists a coarse moduli space $\overline{M}_{g,n}$ for stable n-pointed curves of arithmetic genus g. It is a normal projective variety of dimension $3g - 3 + n$.

Stabilization and contraction work similarly in $\overline{M}_{g,n}$, and the construction of the space is still the one sketched in 1.4. The description of the boundary is a little more complicated since the curves are no longer trees in general: there can be curves of so-called noncompact type, as for example the last curve drawn in 1.2.2. Also, the irreducible components of a stable curve can have ordinary double points, and the locus of irreducible curves with double points forms a new boundary divisor. Just like the boundary divisors we saw in genus zero (cf. 1.5.9), it is the image of a gluing map, as for example $\overline{M}_{0,3} \to \overline{M}_{1,1}$, which takes a 3-pointed rational curve and glues it to itself at two of the marked points, producing a curve with one double point.

1.6.6 Intersection theory on $\overline{M}_{g,n}$ and Witten's conjecture. On $\overline{M}_{g,n}$ there are other tautological classes: since the marked points of an n-pointed curve C are never singular points, each marked point p_i has a well-defined cotangent space. These cotangent lines vary algebraically with the marked points, giving n line bundles on $\overline{M}_{g,n}$. Their first Chern classes are the *psi classes* ψ_i. The forgetful maps relate psi classes to the kappa classes, and in fact every top intersection of psi classes on $\overline{M}_{g,n}$ can be expressed in terms of top intersections of kappa classes on $\overline{M}_g = \overline{M}_{g,0}$ ($g \geq 2$), and vice versa. Considerable interest in the spaces $\overline{M}_{g,n}$ was aroused by the discovery by Witten [86] of a deep connection to string theory and 2D gravity. Based on this connection he conjectured that the generating function for the intersection numbers were governed by the KdV equations (certain partial differential equations dating back to 19th century dynamics). The conjecture was proved by Kontsevich [56], following the reasoning outlined by Witten, and as a consequence, all intersection numbers of psi classes (and kappa classes) on $\overline{M}_{g,n}$ can be computed from the KdV equations.

Exercises

Rational normal curves

1. Recall that a (parametrized) rational normal curve in $\mathbb{P}^d$ is a map

$$\varphi : \mathbb{P}^1 \to \mathbb{P}^d$$

given by $d+1$ linearly independent binary forms of degree d (i.e., they form a basis for $H^0(\mathbb{P}^1, \mathcal{O}(d))$). Special cases: for $d = 2$ we are talking about parametrized smooth conics in $\mathbb{P}^2$, and for $d = 3$ we get parametrized twisted cubics in $\mathbb{P}^3$.

Show that any $d+1$ points on a rational normal curve are independent. *Hint:* the curve is projectively equivalent to (the closure of) the image of $t \mapsto [1 : t : \cdots : t^{d-1} : t^d]$. Plug in $d+1$ distinct points $t_0, \ldots, t_d$. The image points are dependent if and only if the van der Monde determinant $|t_i^j|_{0 \leq i,j \leq d}$ vanishes; this happens if and only if two of its rows coincide.

(For $d = 2$, this is just to say that three points on a smooth conic C are not collinear. By Bézout, if a line intersected C in three points, C would have to contain the line, so the statement is equivalent to saying that C is smooth.)

2. Let there be given a $(d+1)$-tuple of points in $\mathbb{P}^1$ with coordinates $[a_i : b_i]$, $i = 0, \ldots, d$. Assume none of these points are $[1 : 0]$ or $[0 : 1]$. Consider the homogeneous degree-$(d+1)$ form $F(s,t) := \prod_{i=0}^{d}(b_i s - a_i t)$, which vanishes precisely on the $d+1$ points.

   (i) Show that the degree d-forms $R_i := \frac{F}{b_i s - a_i t}$ form a basis for $H^0(\mathbb{P}^1, \mathcal{O}(d))$. Hence the R_i define a rational normal curve. Observe that the point $[a_i : b_i]$ maps to the ith coordinate point $[0 : \cdots : 0 : 1 : 0 : \cdots : 0]$.

   (ii) Show that the point $[0 : 1]$ maps to $[\frac{1}{a_0} : \cdots : \frac{1}{a_d}]$ and $[1 : 0]$ maps to $[\frac{1}{b_0} : \cdots : \frac{1}{b_d}]$. Show that by varying the initial $(d+1)$-tuple, any pair of points in $\mathbb{P}^d$ not on the coordinate planes can be the image of $[1 : 0]$ and $[0 : 1]$ in the above construction.

3. Show that through any $d+3$ general points in $\mathbb{P}^d$ there is a unique rational normal curve.

   (As special cases: through 5 general points in $\mathbb{P}^2$ there is a unique smooth conic, and through 6 general points in $\mathbb{P}^3$ there is a unique twisted cubic.)

4. Given a general $(d+3)$-tuple $\{p_1, \ldots, p_{d+3}\}$ of points in $\mathbb{P}^d$, let $\varphi : \mathbb{P}^1 \to \mathbb{P}^d$ be the unique rational normal curve through these points, and let $q_i := \varphi^{-1}(p_i)$ denote the inverse images of the points. Since the p_i are distinct, also the q_i are distinct, so they form a $(d+3)$-tuple of points in $\mathbb{P}^1$, hence yield a point in $M_{0,d+3}$. Show that two general $(d+3)$-tuples of points in $\mathbb{P}^d$ are projectively equivalent if and only if the corresponding $(d+3)$-tuples in $\mathbb{P}^1$ are projectively equivalent. (This in turn is determined by d cross ratios, cf. 0.1.8.)

The del Pezzo surface S_5

5. Consider the pencil of plane conics through the four coordinate points

$$\begin{bmatrix}1\\0\\0\end{bmatrix}, \begin{bmatrix}0\\1\\0\end{bmatrix}, \begin{bmatrix}0\\0\\1\end{bmatrix}, \begin{bmatrix}1\\1\\1\end{bmatrix}.$$

   Specifically it is the pencil $sX(Y-Z) + tZ(X-Y)$, $\left[\begin{smallmatrix}s\\t\end{smallmatrix}\right] \in \mathbb{P}^1$.

   (i) Describe the three singular members of the pencil in terms of the four coordinate points.

(ii) The linear system defines a rational map $\mathbb{P}^2 \dashrightarrow \mathbb{P}^1$, sending a point $P \in \mathbb{P}^2$ to the point $\left[\begin{smallmatrix} s \\ t \end{smallmatrix}\right]$ corresponding to the conic through the five points. Write out the map explicitly in coordinates.

(iii) Blow up the four base points and check that this resolves the map. The blown-up surface is the del Pezzo surface S_5.

(iv) For each base point, show that there is exactly one conic in the pencil with a given tangent direction, and conclude that the four exceptional divisors provide four (disjoint) sections to the resolved map $S_5 \to \mathbb{P}^1$.

(v) The resolved map $S_5 \to \mathbb{P}^1$ with its four sections is a family of stable 4-pointed curves. Show that it is naturally identified with the universal family $\overline{M}_{0,5} \simeq \overline{U}_{0,4} \to \overline{M}_{0,4}$. In particular, compare the boundary divisors of $\overline{M}_{0,5}$ (see 1.4.1 and 1.5.7) with the 10 distinguished divisors on S_5: the strict transforms of the six lines pairing the four base points, together with the four exceptional divisors.

(vi) Show that the 10 distinguished divisors intersect in 15 points. These correspond to the codimension-2 boundary strata of $\overline{M}_{0,5}$. Show that each divisor contains three of these points and through each of the points there are two divisors. (This is what classically is called a $(10_3, 15_2)$ configuration.)

6. *The 10 lines on the del Pezzo surface* S_5. The linear system of cubics through the four points of the previous exercise defines a rational map $\mathbb{P}^2 \dashrightarrow \mathbb{P}^5$, which embeds the blown-up surface S_5 as a degree-5 surface in $\mathbb{P}^5$ (this is really what is called the del Pezzo surface). Show that the 10 distinguished divisors are exactly the lines contained in $S_5 \subset \mathbb{P}^5$. (See Beauville [4], Ch. IV for more help.)

7. *Kapranov's construction for* $n = 5$. (i) Show that for each smooth 5-pointed curve C there is a unique degree-2 embedding $C \to \mathbb{P}^2$ mapping the first four marked points to the coordinate points. Hence the image of the fifth marked point determines a point in $\mathbb{P}^2$.

(ii) Show more generally that for any family of stable 5-pointed curves $\mathfrak{X}/B$ there is a unique morphism $\mathfrak{X} \to \mathbb{P}^2$ such that the first four sections map constantly to the coordinate points, and conclude that there is induced a morphism $B \to \mathbb{P}^2$ defined by the fifth section. In particular we get a morphism $\overline{M}_{0,5} \to \mathbb{P}^2$.

(iii) Show that this morphism is the blowup of $\mathbb{P}^2$ in the four coordinate points, hence giving another viewpoint on the previous exercises.

8. *Yet another construction of the universal family of* 4*-pointed curves.* Consider a pencil of lines through a point $P \in \mathbb{P}^2$ (say the pencil $s(X-Y)+t(Y-Z)$ of

lines through $P = \begin{bmatrix} 1 \\ 1 \\ 1 \end{bmatrix}$). Resolve the corresponding rational map $\mathbb{P}^2 \dashrightarrow \mathbb{P}^1$ by blowing up in P, to get a family of rational curves $\widetilde{\mathbb{P}}^2 \to \mathbb{P}^1$. The exceptional divisor E gives a first section to the family. Now take three general lines (say the coordinate lines, equations X, Y, and Z): these define another three sections, which intersect pairwise in a point but are disjoint from the section E. Blow up these three points to get a family of stable 4-pointed curves. Show that this is the universal family.

9. Show that the number of boundary divisors of $\overline{M}_{0,n}$ is $2^{n-1} - n - 1$.

Chapter 2

Stable Maps

The definition of a stable map is given in Section 2.3. In the first two sections we dwell on a heuristic discussion. We introduce, in order of sophistication, various parameter spaces of maps, culminating with the statement of the existence of the important Kontsevich moduli spaces $\overline{M}_{0,n}(\mathbb{P}^r, d)$ and their basic properties.

2.1 Maps $\mathbb{P}^1 \to \mathbb{P}^r$

We now turn to our main object of study: rational curves in projective space. The characteristic property of an irreducible rational curve is that it can be parametrized by the projective line; the maps $\mu : \mathbb{P}^1 \to \mathbb{P}^r$ therefore deserve special attention.

Definition. By the *degree* of a map $\mu : \mathbb{P}^1 \to \mathbb{P}^r$ we mean the degree of the direct image cycle $\mu_*[\mathbb{P}^1]$. In particular, a constant map has degree zero.

In other words, if $e \geq 1$ is the degree of the image curve (with reduced scheme structure), and k denotes the degree of the field extension corresponding to the map, then the degree of the map is $k \cdot e$. Note that, except for the case in which the image curve is a straight line, the definition above differs from the usual definition, given just by the degree of the field extension.

2.1.1 The space of parametrizations. To give a map $\mu : \mathbb{P}^1 \to \mathbb{P}^r$ of degree d is to specify, up to a constant factor, $r + 1$ binary forms of degree d, which are not allowed to vanish simultaneously at any point. This condition defines a Zariski open subset

$$W(r, d) \subset \mathbb{P}\Big(\bigoplus_{i=0}^{r} H^0(\mathcal{O}_{\mathbb{P}^1}(d))\Big).$$

The dimension of $W(r, d)$ is $rd + r + d$. Indeed, there are $(r + 1)(d + 1)$ degrees of freedom for choosing the binary forms; subtract 1 because two $(r+1)$-tuples define the same map if they differ by a constant factor.

The space $W(r, d)$ is equipped with an obvious family of maps,

$$\begin{array}{ccc} W(r,d) \times \mathbb{P}^1 & \longrightarrow & \mathbb{P}^r \\ \downarrow & & \\ W(r,d) & & \end{array}$$

where the horizontal arrow maps (μ, x) to $\mu(x)$. In fact this family is the universal family: any other family $B \times \mathbb{P}^1 \to \mathbb{P}^r$ of maps of degree d is induced from it by pullback along a unique morphism $B \to W(r, d)$. In other words, $W(r, d)$ is a fine moduli space for maps $\mathbb{P}^1 \to \mathbb{P}^r$ of degree d.

In what follows it is convenient to assume $d \geq 1$.

We use the term *immersion* for a morphism whose tangent map is injective at any point.

2.1.2 Lemma. *The locus $W^\circ(r, d) \subseteq W(r, d)$ consisting of immersions is open. For $d = 1$, $W^\circ(r, 1)$ is equal to $W(r, 1)$; for $d \geq 2$, its complement is of codimension $r - 1$.*

Proof. A linear map has no ramification, so we can assume $d \geq 2$. Consider the closed subset $\Sigma := \{(\mu, x) \in W(r, d) \times \mathbb{P}^1 \mid D\mu_x = 0\}$. Then $R := W(r, d) \setminus W^\circ(r, d)$ is the image of the projection $\Sigma \to W(r, d)$, and hence closed. This morphism is finite, since a given map $\mu \in W(r, d)$ has only a finite number of ramification points. Let us compute the dimension of the fibers of the projection $\Sigma \to \mathbb{P}^1$. It is sufficient to look at the point $[1 : 0]$. In affine coordinates, the map is given by $r + 1$ polynomials $f_0, \ldots, f_r$ of degree $\leq d$, say $f_k(t) = a_{k0} + a_{k1}t + \cdots + a_{kd}t^d$. We can assume that f_0 does not vanish at $t = 0$ (this is, $a_{00} \neq 0$). In an affine neighborhood the map is then given by $t \mapsto (f_1/f_0, \ldots, f_r/f_0)$. Its derivative at $t = 0$ is the vector of the derivatives of f_k/f_0, evaluated at $t = 0$, which gives

$$\frac{a_{00}a_{k1} - a_{01}a_{k0}}{a_{00}^2}.$$

Since $a_{00} \neq 0$, the vanishing of the derivative amounts to r independent conditions in the a_{ij}'s. That is, the fiber of $\Sigma \to \mathbb{P}^1$ is of dimension $rd + d$. Therefore Σ, and consequently R, has dimension $rd + d + 1$, which is equivalent to the claimed codimension. □

2.1.3 Birational maps and multiple covers. Denote by $W^*(r, d)$ the locus in $W(r, d)$ constituted by maps that are birational onto their image. An immersion is always birational onto its image, so we have $W^\circ(r, d) \subset W^*(r, d) \subset W(r, d)$. For $d = 1$, we have $W^\circ(r, 1) = W^*(r, 1) = W(r, 1)$. For $d \geq 2$, the maps in the complement of $W^*(r,d)$ are the *multiple covers*, i.e., maps $\mathbb{P}^1 \to \mathbb{P}^r$ such that the field extension corresponding to source and target is of degree at least 2. Every multiple cover factorizes as $\mathbb{P}^1 \xrightarrow{\rho} \mathbb{P}^1 \xrightarrow{\psi} \mathbb{P}^r$, where ρ is a multiple cover of the line, and ψ is a birational map onto its image. The factorization is not unique because we can always insert an automorphism of $\mathbb{P}^1$ followed by its inverse: if $\mu = \psi \circ \rho$ is a factorization, then for each $\phi \in \mathrm{Aut}(\mathbb{P}^1)$ we get another factorization $\mu = (\psi \circ \phi^{-1}) \circ (\phi \circ \rho)$. Conversely, if $\mu = \psi \circ \rho = \widetilde{\psi} \circ \widetilde{\rho}$, it follows that ψ, $\widetilde{\psi}$ have the same image line, so it is licit to put $\phi := \widetilde{\psi}^{-1}\psi \in \mathrm{Aut}(\mathbb{P}^1)$. Thus one may write $\widetilde{\psi} = \psi \circ \phi^{-1}$ and $\widetilde{\rho} = \phi \circ \rho$.

2.1.4 Lemma. *The locus in $W(r,d)$ of d-fold covers of a line is closed of codimension $(r-1)(d-1)$.*

Proof. Such a map factorizes as $\mathbb{P}^1 \xrightarrow{\rho} \mathbb{P}^1 \xrightarrow{\psi} \mathbb{P}^r$, where $\rho \in W(1,d)$ and $\psi \in W(r,1)$. We then get a natural morphism

$$\begin{array}{ccc} W(1,d) \times W(r,1) & \longrightarrow & W(r,d) \\ (\rho,\psi) & \longmapsto & \psi \circ \rho, \end{array}$$

whose image is the locus we want. The dimension of the product is $(2d+1) + (2r+1)$. On the other hand, the space $W(r,d)$ has dimension $rd+r+d$, and the fibers are all of dimension $3 = \dim \mathrm{Aut}(\mathbb{P}^1)$. Therefore the image is of codimension $(rd+r+d)+3-(2d+1)-(2r+1) = (r-1)(d-1)$. Heuristically, it is closed because a multiple cover of a line cannot degenerate (in $W(r,d)$) into any other type of parametrization. Indeed, the limit map continues having a line as its image. To put this on solid ground, set for short $M = W(r,d)$. Let G denote the Grassmannian of lines $L \subseteq \mathbb{P}^r$. Form the correspondence $V = \{(\mu, L) \in M \times G \mid \mu(\mathbb{P}^1) \subseteq L\}$. The fiber of V over the line L given by $x_2 = \cdots = x_r = 0$ is isomorphic to $W(1,d)$. Hence V is irreducible and its dimension is equal to $\dim W(1,d) + \dim G = 2d+1+2(r-1)$. Since G is projective and the projection $V \to M$ is injective, it follows that its image is closed and of codimension $rd+r+d-2d-2r+1$ as asserted. □

2.1.5 Lemma. *Suppose d is even. Then the closure in $W(r,d)$ of the locus of double covers of curves of degree $d/2$ is of codimension $(r+1)d/2-2$.*

Proof. This follows from an argument similar to that of the preceding proof, looking now at the morphism

$$\begin{array}{ccc} W(1,2) \times W^*(r,d/2) & \longrightarrow & W(r,d) \\ (\rho,\psi) & \longmapsto & \psi \circ \rho. \end{array}$$

This time the image is only constructible, since the birational factor ψ can degenerate into some multiple cover, thus jumping to another type of factorization. □

2.1.6 Example. (*Parametrizations of quartics*) The space of all the parametrizations of plane quartics $W(2,4)$ is of dimension 14. Inside it, the locus of 4-fold lines has dimension 11, while the locus of double conics has dimension 10.

For $\mathbb{P}^3$ we have $\dim W(3,4) = 19$, and in this space, the 4-fold covered lines constitute a family of dimension 13; by coincidence this number is also the dimension of the locus of double conics.

Now in $\mathbb{P}^4$, we find that $\dim W(4,4) = 24$, but here the dimension of the locus of 4-fold lines (15) is surpassed by the family of double conics (dimension 16).

The two cases treated in the previous lemmas are in fact the extreme cases, in the sense that any other possible factorization of d into two natural numbers yields higher codimension (cf. Pandharipande [69], Lemma 2.1.1). Precisely, for $d = k \cdot e$, consider k-sheeted covers of curves of degree e. Then the locus in $W(r, d)$ of such maps is of codimension $(k-1)(e(r+1)-2)$. This follows from arguments similar to those of the two lemmas, applied to the map

$$W(1,k) \times W(r,e) \to W(r,d).$$

We summarize this discussion in the following proposition.

2.1.7 Proposition. *The locus $W^*(r,d) \subseteq W(r,d)$ formed by the maps that are birational onto their image is open. If $d \geq 2$, then its complement is of codimension at least*

$$\min\big\{(r-1)(d-1)\,,\ (r+1)d/2 - 2\big\}.$$

In particular, if $r \geq 2$, $W^*(r,d)$ is dense in $W(r,d)$. □

To establish openness, set for short $W = W(r,d)$. Put

$$D = \{(\mu, x) \in W \times \mathbb{P}^1 \mid \exists y \neq x \text{ with } \mu x = \mu y \text{ or } \mu' x = 0\},$$

the double points together with ramification locus. It is closed. If $\mu \in W$ is birational, then the fiber D_μ is finite, and vice versa. By semicontinuity of fiber dimension of $D \to W$, it follows that W^* is open.

2.1.8 Drawbacks of $W(r,d)$. As a tool for describing families of rational curves, $W(r,d)$ has some serious drawbacks. While by definition, every rational curve admits a parametrization, it is not true that every family of rational curves admits a family of parametrizations from *one and the same* $\mathbb{P}^1$; cf. Example 2.1.10 below.

Another problem with $W(r,d)$ is redundancy: reparametrizations of the same rational curve in $\mathbb{P}^r$ are considered distinct objects. We need to pass to the quotient of this equivalence.

An additional defect of $W(r,d)$ is that it is not compact (complete).

Let us analyze the first drawback.

2.1.9 Definition. A *family of maps* of smooth rational curves is a diagram

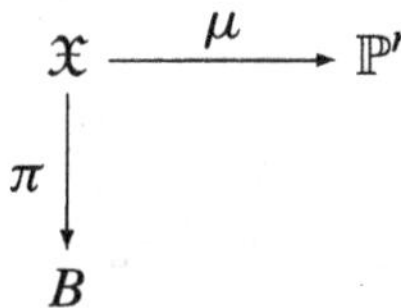

where π is a flat family with geometric fibers isomorphic to $\mathbb{P}^1$. In this way, for each $b \in B$, the map μ restricted to the fiber, $\mu_b : \mathfrak{X}_b \to \mathbb{P}^r$, is a map from a smooth rational curve. It can be shown that all μ_b have the same degree. Assuming that $\mathfrak{X} \to B$ is the projectivization of a vector bundle E over B, the argument is roughly this. Set $L = \mu^*\mathcal{O}(1)$. Then we have $L = \mathcal{O}_E(r) \otimes \pi^*\beta$, for some line bundle β over B (cf. [44, 12.5, p. 291]). It follows that $\mu_b^*\mathcal{O}(1) = \mathcal{O}_{E_b}(r)$ for all $b \in B$.

Often, for typographic convenience, we will indicate such a family by $\mathfrak{X} \to B \times \mathbb{P}^r$, understanding that B is the base variety of the family.

2.1.10 Example. Let $\mathbb{P}^2 \dashrightarrow \mathbb{P}^1$ be the projection given by $[x : y : z] \mapsto [x : y]$. The map is resolved by the blowup $\mu : \widetilde{\mathbb{P}}^2 \to \mathbb{P}^2$ in the point $[0 : 0 : 1]$. Indeed, $\widetilde{\mathbb{P}}^2$ is the closure in $\mathbb{P}^2 \times \mathbb{P}^1$ of the graph of the projection. Let $\pi : \widetilde{\mathbb{P}}^2 \to \mathbb{P}^1$ be the resolved map. Then we have a diagram

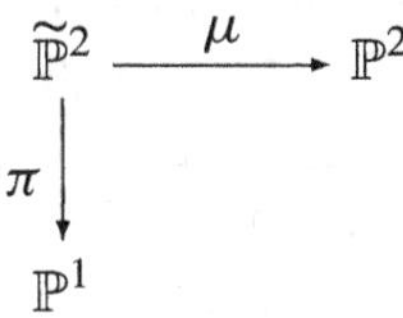

which is a family of irreducible maps of degree 1, because all the fibers are isomorphic to $\mathbb{P}^1$ and they are mapped to lines through $[0 : 0 : 1] \in \mathbb{P}^2$. However, the family is not trivial: $\widetilde{\mathbb{P}}^2 \simeq \mathbb{P}(\mathcal{O} \oplus \mathcal{O}(1))$, so there is no continuous way to identify all the fibers with one and the same $\mathbb{P}^1$.

2.1.11 Isomorphisms and automorphisms. An *isomorphism* between maps $\mu : C \to \mathbb{P}^r$ and $\mu' : C' \to \mathbb{P}^r$ is an isomorphism $\phi : C \xrightarrow{\sim} C'$ that turns this diagram commutative:

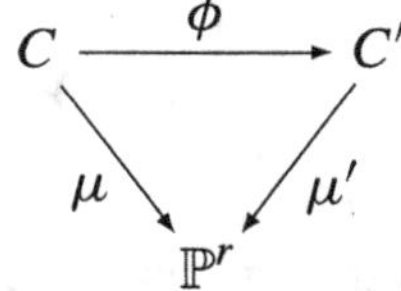

We then say that μ and μ' are *isomorphic* maps. Correspondingly, an *automorphism* of a map $\mu : C \to \mathbb{P}^r$ is an automorphism $\phi : C \xrightarrow{\sim} C$ such that $\mu \circ \phi = \mu$. The notion of isomorphism for families is defined in the evident way.

When classifying algebro-geometric objects up to isomorphism, as a rule of thumb, if every object is automorphism-free (i.e., has a trivial group of automorphisms), then in general the existence of a fine moduli space can be expected (but see however 0.2.10); if each object has a finite group of automorphisms, one expects to get only a coarse moduli space (but see however Exercise 15 on page 19); and if some object has an infinite automorphism group, not even a coarse moduli space can be expected.

As a response to the second drawback of $W(r, d)$, we want to classify maps $\mathbb{P}^1 \to \mathbb{P}^r$ (of degree $d \geq 1$) up to isomorphism, i.e., provide the quotient set

$$W(r, d) \big/ \operatorname{Aut}(\mathbb{P}^1)$$

with the structure of a variety. In view of the above discussion, the following lemma makes the existence of a coarse moduli space plausible.

2.1.12 Lemma. *Let $\mu : \mathbb{P}^1 \to \mathbb{P}^r$ be a nonconstant map. Then there is only a finite number of automorphisms $\phi : \mathbb{P}^1 \xrightarrow{\sim} \mathbb{P}^1$ such that $\mu = \mu \circ \phi$. If μ is birational onto its image, then* $\operatorname{Aut}(\mu)$ *is trivial.*

Proof. Let K be the function field of the image curve $\mu(\mathbb{P}^1) \subset \mathbb{P}^r$, and let L be the function field of $\mathbb{P}^1$. Then the automorphism group of μ is naturally identified with the group of automorphisms of L compatible with the finite field extension $K \hookrightarrow L$, and this group is known to be finite. Saying that μ is birational onto its image is just to say $L = K$, hence the automorphism group is trivial. □

2.1.13 Example. The double cover

$$\begin{array}{ccc} \mathbb{P}^1 & \to & \mathbb{P}^2 \\ [x : y] & \mapsto & [x^2 : y^2 : 0] \end{array}$$

admits a unique nontrivial automorphism, namely $[x : y] \mapsto [-x : y]$.

2.1.14 The moduli space of maps $\mathbb{P}^1 \to \mathbb{P}^r$. The preceding remarks show that the open set $W^*(r, d) \subset W(r, d)$ of maps that are birational onto their image is precisely the set of automorphism-free maps. It is not surprising then to learn that there exists a fine moduli space

$$M^*_{0,0}(\mathbb{P}^r, d) \simeq W^*(r, d) / \operatorname{Aut}(\mathbb{P}^1);$$

in fact this is the geometric quotient in the sense of Mumford [62], (cf. Kollár [55], p. 105). For an interesting introductory discussion and guide to the literature on the problem of constructing quotient spaces in algebraic geometry, the reader can consult Esteves [25] or Newstead [66]; otherwise, the standard reference is [62].

Assuming $d \geq 1$, the maps in the complement of $W^*(r, d)$ are precisely the multiple-cover maps. Including those maps, what we get is only a coarse moduli space

$$M_{0,0}(\mathbb{P}^r, d) \simeq W(r, d)/\operatorname{Aut}(\mathbb{P}^1).$$

We will not substantiate these statements any further, and the statements themselves will we subsumed in Theorem 2.3.2, where we consider the Kontsevich compactification (and finally get rid of the third drawback of $W(r, d)$).

In the notation for these spaces, the first subscript indicates that we are considering curves of genus 0; the second subscript indicates that we do not (yet) put marks on the source curve; we will do that in Section 2.3, and also make sense of the case $d = 0$.

The dimension of $M_{0,0}(\mathbb{P}^r, d)$ is $rd + r + d - 3$, as expected from the following dimension count. There is a morphism $W(r, d) \to M_{0,0}(\mathbb{P}^r, d)$ (the classifying map). The generic fiber of this morphism is $\operatorname{Aut}(\mathbb{P}^1)$. Therefore, the dimension of $M_{0,0}(\mathbb{P}^r, d)$ must be

$$\dim M_{0,0}(\mathbb{P}^r, d) = \dim W(r, d) - \dim \operatorname{Aut}(\mathbb{P}^1) = rd + r + d - 3.$$

Note that for for $r \geq 2$ (and $d \geq 1$), Lemma 2.1.7 implies that $M^*_{0,0}(\mathbb{P}^r, d)$ is dense in $M_{0,0}(\mathbb{P}^r, d)$.

2.1.15 Example: $d = 1$**.** Each map of degree one is a parametrization of a line. But we pass to the quotient in order to identify reparametrizations, so the equivalence class of a map can be perfectly matched to its image line. Hence $M_{0,0}(\mathbb{P}^r, 1)$ must be the Grassmannian $\operatorname{Gr}(1, \mathbb{P}^r)$. To make the bijection explicit, note that a linear map $\mathbb{P}^1 \to \mathbb{P}^r$ is given by an $(r + 1) \times 2$ matrix A of rank 2. The map is given in coordinates by

$$\begin{bmatrix} x_0 \\ x_1 \end{bmatrix} \mapsto A \cdot \begin{bmatrix} x_0 \\ x_1 \end{bmatrix}.$$

Now we have to mod out by reparametrizations. Each linear reparametrization can be written as $\left[\begin{smallmatrix} x_0 \\ x_1 \end{smallmatrix}\right] = \left[\begin{smallmatrix} a & b \\ c & d \end{smallmatrix}\right]\left[\begin{smallmatrix} y_0 \\ y_1 \end{smallmatrix}\right]$. It follows that the group $\operatorname{Aut}(\mathbb{P}^1)$ acts on the space of matrices A by multiplication on the right, that is, by column operations. But the variety of $(r+1) \times 2$ matrices of rank 2 modulo column operations is precisely the Grassmannian $\operatorname{Gr}(1, \mathbb{P}^r)$.

Note in particular that in this case, the space is actually compact. Note also that every linear map is birational (onto its image), and therefore is free from automorphisms. In other words, in the present (rather special!) case we have

$$M^*_{0,0}(\mathbb{P}^r, 1) = M_{0,0}(\mathbb{P}^r, 1) = \overline{M}_{0,0}(\mathbb{P}^r, 1).$$

For $d \geq 2$ it is impossible to avoid automorphisms: $M^*_{0,0}(\mathbb{P}^r, d) \neq M_{0,0}(\mathbb{P}^r, d)$.

A natural idea to suppress automorphisms is simply to put marks on the source curve, and require that the automorphisms respect this structure as well. If every source curve has three marked points or more, clearly there can be no automorphisms left. In fact, the moduli space is easy to describe in this situation:

2.1.16 Proposition. *For each $n \geq 3$ there is a fine moduli space $M_{0,n}(\mathbb{P}^r, d)$ for isomorphism classes of n-pointed maps $\mathbb{P}^1 \to \mathbb{P}^r$ of degree d, namely*

$$M_{0,n}(\mathbb{P}^r, d) = M_{0,n} \times W(r, d).$$

In particular, $M_{0,n}(\mathbb{P}^r, d)$ is a smooth variety; it is an open set in the "linear space" $\mathbb{A}^{n-3} \times \mathbb{P}^{rd+r+d}$.

Proof. Let us show that the following family has the universal property:

$$\begin{array}{ccc} M_{0,n} \times W(r, d) \times \mathbb{P}^1 & \xrightarrow{\mu} & \mathbb{P}^r \\ \sigma_i \uparrow \cdots \uparrow \; \downarrow & & \\ M_{0,n} \times W(r, d), & & \end{array}$$

where the n sections σ_i are those of $M_{0,n}$ (cf. 1.1.5), with the first three sections rigidified as $0, 1, \infty$, in this order.

Let $\mathfrak{X} \to B \times \mathbb{P}^r$ be an arbitrary family of n-pointed maps of degree d. We must show that there exists a unique morphism $B \to M_{0,n} \times W(r, d)$ inducing $\mathfrak{X}$ as the pullback of the claimed universal family. Now, since there are at least three disjoint sections, we know that $\mathfrak{X}/B$ is isomorphic to the trivial family $B \times \mathbb{P}^1 \to B$ (cf. 1.1.3). There are infinitely many such isomorphisms, but only one that identifies the first three sections with $0, 1, \infty$, in this order. The n sections turn $\mathfrak{X}/B$ into a family of n-pointed smooth curves. By the universal property of $M_{0,n}$, there is a unique morphism $B \to M_{0,n}$ inducing $\mathfrak{X}$ from the universal family $M_{0,n} \times \mathbb{P}^1 \to M_{0,n}$ in a way compatible with the n sections (cf. 1.1.5).

On the other hand, the universal property of $W(r, d)$ (cf. 2.1.1) ensures that our family $B \times \mathbb{P}^1 \to \mathbb{P}^r$ is induced form the universal family $W(r, d) \times \mathbb{P}^1 \to \mathbb{P}^r$ via a unique morphism $B \to W(r, d)$. Combining the two morphisms we obtain $B \to M_{0,n} \times W(r, d)$ inducing $\mathfrak{X}$, as desired. □

2.2 1-parameter families

In this section we experiment with 1-parameter families of maps $\mathbb{P}^1 \to \mathbb{P}^r$, i.e., families $\mathfrak{X} \to B \times \mathbb{P}^r$ whose base B is a curve, for example an open set in $\mathbb{A}^1$.

The idea is that each such family has a classifying map $B \to M_{0,0}(\mathbb{P}^r, d)$ and thus defines an arc in the moduli space. By studying the natural limits of such families we get a picture of what kinds of objects we need to take in to compactify $M_{0,0}(\mathbb{P}^r, d)$. It will quickly be apparent that (for $d \geq 2$) it is necessary to include maps with reducible source.[1] Our analysis will suggest the definition of Kontsevich stability, given in the next section. Needless to say, our discussion is far from being a proof of the fact that the notion actually leads to a compact and separated space.

2.2.1 Example. We start out with the pencil of conics in $\mathbb{P}^2$ given by the family of equations $XY - bZ^2$, with parameter $b \in B := \mathbb{A}^1$. All the members of the family are smooth conics except the special member $b = 0$, which is the pair of lines XY:

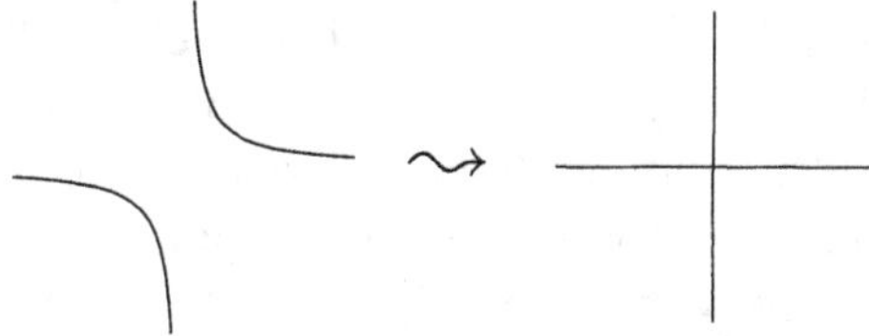

We can describe the pencil as a family of parametrizations, except for the special member. To this end, we consider the rational map

$$\begin{array}{ccc} B \times \mathbb{P}^1 & \dashrightarrow & \mathbb{P}^2 \\ \left(b, \begin{bmatrix} s \\ t \end{bmatrix}\right) & \longmapsto & \begin{bmatrix} bs^2 \\ t^2 \\ st \end{bmatrix} \end{array}$$

as a family of maps $\mathbb{P}^1 \to \mathbb{P}^2$ indexed by the parameter $b \in B = \mathbb{A}^1$. Note that the map is given by three sections of the line bundle $\mathcal{O}(2)$ on $B \times \mathbb{P}^1$. The map is not defined at $\big(0, [1:0]\big)$, that is, we are talking about a base point of the corresponding linear system. The central fiber, except for the base point, is mapped to the line $X = 0$. But we know from the theory of surfaces (as exposed for example in Beauville [4]) that it is possible to resolve the indeterminacy of the map by blowing up the base point. In the affine chart given by $s = 1$ (the only interesting chart in this case), the map becomes

$$(b, t) \mapsto [b : t^2 : t].$$

The ideal of the base locus is $\langle b, t\rangle$, so the blowup sits inside $\mathbb{A}^2 \times \mathbb{P}^1$ with coordinates $(b, t) \times [b_1 : t_1]$ as the subvariety given by the equation $bt_1 = tb_1$. The interesting chart of the blowup is that with $t_1 = 1$. Substituting $b = tb_1$ we obtain the "total transform" of the map,

$$[tb_1 : t^2 : t],$$

[1]Or, include in the boundary objects that do not correspond to maps cf. [77].

but to get the resolved map we must divide out by the factor corresponding to the exceptional divisor (equation $t = 0$), getting

$$[b_1 : t : 1].$$

We are interested in the values in the fiber over $b = 0$. Here the source curve is the union of the strict transform F of the fiber ($b_1 = 0$) and the exceptional divisor E (given by $t = 0$). For $b_1 = 0$, the map is $t \mapsto [0 : t : 1]$, whose image is the straight line with equation X. For $t = 0$ we get $b_1 \mapsto [b_1 : 0 : 1]$, which gives the line Y. In other words, the natural limit of this family of degree-2 maps is the "union" of two maps of degree 1.

The example indicates that if we want a compact moduli space, we must include some maps of type $C \to \mathbb{P}^r$ where the source is reducible. The reducible curves appearing as a result of blowing up are always trees of rational curves.

Now this limit is not the only possible one: we could for example get another limit simply by performing yet another blowup at a point of the central fiber. This curve would then be a curve with three twigs, one of which would contract to a single point in $\mathbb{P}^2$. This is, the limit map would then be the "union" of three maps, of degrees 1, 1, and 0.

It would be very bad to allow a 1-parameter family to have various different limits: this would amount to a moduli space that would not be separated! (Recall the valuative criterion for separatedness; see [44, Ch. II, §4].)

An obvious idea to avoid such pathologies is simply to interdict twigs of degree zero in the source curve, considering them artificial. This way we might be tempted to invite *temporarily* only the following maps to form the boundary of the moduli space of maps of degree d: all maps $\mu : C \to \mathbb{P}^r$ where C is a tree of smooth rational curves such that the restriction of μ to each twig is a map of positive degree and the sum of the degrees is d.

However, even in this way we would not yet have invited sufficiently many maps to form the boundary that compactifies the space, as the following example illustrates.

2.2.2 Example. Let us degenerate an irreducible nodal cubic $F = Y^2Z - X^2(X - Z)$ into the union of three concurrent lines given by the polynomial $G = X(X - Y)(X + Y)$. Take the pencil $bF + G$ and let b approach zero:

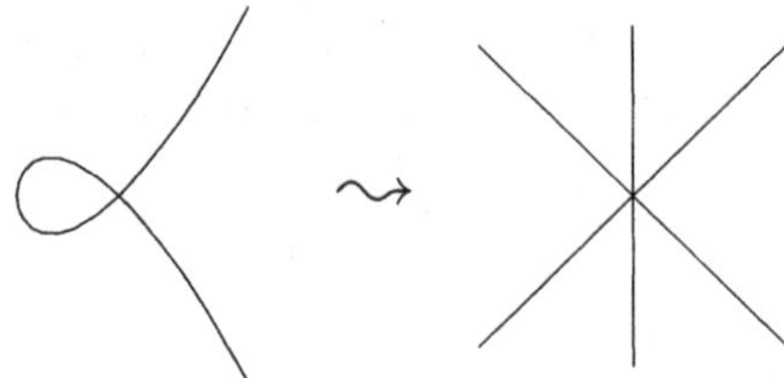

Note that for any value of b, the corresponding cubic in the pencil has $[0:0:1]$ as a singular point. Therefore all nonspecial members of the family are rational curves.

In order to find a family of parametrizations, we intersect with the pencil $sX+tY$ of lines through the point $[0:0:1]$. Each line meets the cubic with multiplicity 2 at the origin; the third point of intersection describes the cubic parametrically. We obtain the rational map

$$\begin{array}{ccc} B\times\mathbb{P}^1 & \dashrightarrow & \mathbb{P}^2 \\ (b,[s:t]) & \longmapsto & \begin{bmatrix} bt(s^2+t^2) \\ -bs(s^2+t^2) \\ t(s^2-t^2+bt^2) \end{bmatrix}. \end{array}$$

We see that this map has certain base points for $b=0$, precisely $t=0$ and $t=\pm s$. The rest of the fiber over $b=0$ is contracted to the point $[0:0:1]$. (There are also two base points in the fiber $b=2$, but they are of no interest here.) Set $s=1$ and consider the map in the corresponding affine chart,

$$(b,t) \longmapsto \begin{bmatrix} bt(1+t^2) \\ -b(1+t^2) \\ t(1-t^2+bt^2) \end{bmatrix}.$$

We must blow up the surface in each of the three base points and determine whether the map is resolved. We treat only the blowup at the point $(b,t)=(0,0)$, which is the simplest. Set $bt_1=tb_1$ and look in the affine chart $t_1=1$. Substituting $b=tb_1$, we obtain

$$\begin{bmatrix} b_1t(1+t^2) \\ -b_1(1+t^2) \\ 1-t^2+b_1t^3 \end{bmatrix},$$

where a factor t was canceled. We are interested in the values in the fiber over $b=0$. Here, the source is the union of the strict transform F of the fiber $(b_1=0)$ and the exceptional divisor E (given by $t=0$). For $b_1=0$, the map is $t\mapsto[0:0:1-t^2]$, constant to the origin. For $t=0$ we get $b_1\mapsto[0:-b_1:1]$, which is the line X.

Blowing up also the two other base points, $t=\pm1$, we see that the map is really resolved, and the new central fiber is of type

```
1 -+--------
1 -+--------
1 -+--------
   0
```

where the vertical twig represents the strict fiber (mapping to the origin $[0:0:1]\in\mathbb{P}^2$), and the horizontal ones are the three exceptional divisors, which are mapped to the lines X, $X+Y$ and $X-Y$, respectively.

In conclusion: we have a limit map whose source has naturally acquired a twig of degree zero. However, it is not possible to blow down (contract) that twig: that would yield a source curve with a triple point, which is the kind of objects meticulously precluded in the Deligne–Mumford–Knudsen compactification! The moral is that we must allow twigs of degree zero, under the condition that they intersect the other twigs in at least three points (and for this reason are unavoidable).

These considerations lead to the notion of Kontsevich stability.

2.3 Kontsevich stable maps

The notion of Kontsevich stability applies to *n-pointed maps*, and combines the structures studied in the previous section with the structure studied in Chapter 1.

There are two good reasons for incorporating the marks in the definition: the first is that even if our primary interest were just maps without marked points, the description of the boundary of $\overline{M}_{0,0}(\mathbb{P}^r, d)$ turns out to have a natural expression in terms of maps of lower degree where the marked points play an important role to compatibilize gluings; see 2.7.3.

Another reason is that we are going to do enumerative geometry, counting maps subject to conditions that are most naturally expressed in terms of the images of the marked points, cf. 2.5.2 and Chapter 3.

Definition. An *n-pointed map* is a morphism $\mu : C \to \mathbb{P}^r$, where C denotes a tree of projective lines with n distinct marked points that are smooth points of C. An *isomorphism* of n-pointed maps $\mu : C \to \mathbb{P}^r$ and $\mu' : C' \to \mathbb{P}^r$ is an isomorphism of the source curves that respects all the structure, i.e., $\phi : C \xrightarrow{\sim} C'$, making these two diagrams commutative:

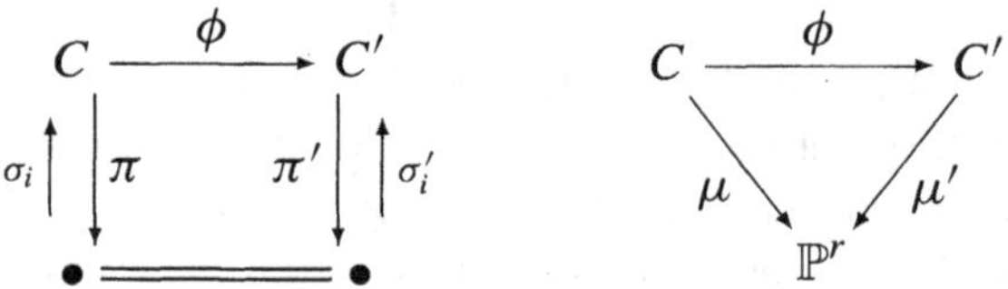

More generally, a family of n-pointed maps is a diagram

$$\begin{array}{ccc} \mathfrak{X} & \xrightarrow{\mu} & \mathbb{P}^r \\ \sigma_i \uparrow \downarrow \pi & & \\ B & & \end{array}$$

where π is a flat family of trees of smooth rational curves, and the σ_i are n disjoint sections that do not meet the singularities of the fibers of π. In this way, for each

$b \in B$, the map μ restricted to the fiber $\mu_b : \mathfrak{X}_b \to \mathbb{P}^r$ is an n-pointed map, with the marked points given by $\sigma_1(b), \ldots, \sigma_n(b)$. The notion of isomorphism of families is defined in the obvious way.

Definition. An n-pointed map $\mu : C \to \mathbb{P}^r$ is called *Kontsevich stable* if any twig mapped to a point is stable as a pointed curve; that is, there must be at least three special points on it. Recall that a special point is either a marked point or a singular point (i.e., a point where the twig intersects another twig). Note that the source curve of a stable map is not necessarily a stable curve. For example every non-constant map $\mathbb{P}^1 \to \mathbb{P}^r$ is stable, but if there are no marked points on it the source curve is not a stable curve.

The reason for this definition is revealed by the following lemma.

2.3.1 Lemma. *An n-pointed map is Kontsevich stable if and only if it has only a finite number of automorphisms.*

Proof. Let μ be a Kontsevich stable map. If its source curve $(C; p_1, \ldots, p_n)$ is stable as an n-pointed rational curve, then there are no automorphisms. If there exists a twig, say E, that is unstable as an n-pointed curve, then by Kontsevich stability, μ is not mapped to a point. Let ϕ be an automorphism of μ. Set $E' = \phi(E)$. We have $\mu_{|E'} \circ \phi_{|E} = \mu_{|E}$. Now Lemma 2.1.12 guarantees that there are only finitely many automorphisms of $\phi_{|E}$.

Conversely, suppose μ is not stable. Then there is an unstable twig E mapping to a point. This twig admits an infinity of automorphisms. Each automorphism of E extends to C declaring it to be the identity on the other twigs. Since the image of $\mu(E)$ is a point, these automorphisms commute with μ, which therefore admits infinitely many automorphisms. □

We state without proof the following existence theorem. While this is a deep result, it is not surprising, in view of the above discussion.

2.3.2 Theorem. (*Cf.* FP-NOTES) *There exists a coarse moduli space $\overline{M}_{0,n}(\mathbb{P}^r, d)$ parametrizing isomorphism classes of Kontsevich stable n-pointed maps of degree d to $\mathbb{P}^r$.* □

The only type of stability considered for maps will be Kontsevich stability. Therefore we will suppress the attribute "Kontsevich" and simply speak of stable maps.

The fundamental properties of the Kontsevich spaces are listed in the following theorem.

2.3.3 Theorem. (*Cf.* FP-NOTES) *$\overline{M}_{0,n}(\mathbb{P}^r, d)$ is a projective normal irreducible variety, and it is locally isomorphic to a quotient of a smooth variety by the action of a finite group. It contains $\overline{M}^*_{0,n}(\mathbb{P}^r, d)$ as a smooth open dense subvariety which is a fine moduli space for maps without automorphisms.* □

Specifying that $\overline{M}_{0,n}(\mathbb{P}^r, d)$ is a projective variety implies that it is separated and complete. In other words, given a 1-parameter family with one member missing, there is exactly one way to complete the family. Thus we have excluded the situation we imagined in Example 2.2.1 of the successive blowups at points in the central fiber.

2.3.4 Remark. We should mention that the moduli space of stable maps has a cleaner description in the language of stacks: the stack version of $\overline{M}_{0,n}(\mathbb{P}^r, d)$ is smooth (technically it is a smooth and proper Deligne–Mumford stack) and possesses a universal family, naturally identified with (the stack version of) $\overline{M}_{0,n+1}(\mathbb{P}^r, d)$. We will briefly comment on this in 2.10.5.

2.3.5 The dimension of $\overline{M}_{0,n}(\mathbb{P}^r, d)$ is

$$\dim \overline{M}_{0,n}(\mathbb{P}^r, d) = (r+1)(d+1) - 1 - 3 + n = rd + r + d + n - 3,$$

as follows from the count made in 2.1.14, together with the observation that each mark increments the dimension by one.

2.4 Idea of the construction of $\overline{M}_{0,n}(\mathbb{P}^r, d)$

2.4.1 The general idea. The Deligne–Mumford–Knudsen spaces $\overline{M}_{0,m}$ play a fundamental role. In fact, $\overline{M}_{0,n}(\mathbb{P}^r, d)$ is the result of gluing together quotients of smooth varieties that are fibrations over open sets of $\overline{M}_{0,m}$, with $m = n + d(r+1)$. For simplicity we restrict ourselves to the case $r = 2$.

2.4.2 Description of an open set in $\overline{M}_{0,n}(\mathbb{P}^2, d)$. Fix three lines ℓ_0, ℓ_1, ℓ_2 in $\mathbb{P}^2$, defined by three independent linear forms $x_0, x_1, x_2 \in H^0(\mathcal{O}_{\mathbb{P}^2}(1))$. We are interested in the open set of $\overline{M}_{0,n}(\mathbb{P}^2, d)$ consisting of all maps transverse to these lines. Precisely, we are considering the open set of maps $\mu : C \to \mathbb{P}^2$ such that *the inverse image of the divisor $\ell_0 + \ell_1 + \ell_2$ consists of* 3d *distinct nonspecial points of* C. Note that the points of the divisor $D_j := \mu^*\ell_j$ are distributed on the twigs in accordance with their degree (2.1): if, for example, μ restricted to a twig has degree d_A, then D_j has d_A points on this twig. Let us denote the points of D_j by the symbols $q_{j1}, \ldots, q_{jd}$,

$$D_j = q_{j1} + \cdots + q_{jd}.$$

Note that the three divisors D_0, D_1, D_2 are linearly equivalent. In fact, they are given by the sections

$$s_0 := \mu^*x_0, \quad s_1 := \mu^*x_1, \quad s_2 := \mu^*x_2$$

of the same line bundle $\mu^*\mathcal{O}_{\mathbb{P}^2}(1)$.

2.4.3 Example. Consider the figure below, of a stable map $\mu : C \to \mathbb{P}^2$ in $\overline{M}_{0,2}(\mathbb{P}^2, 5)$ with three twigs in the source curve.

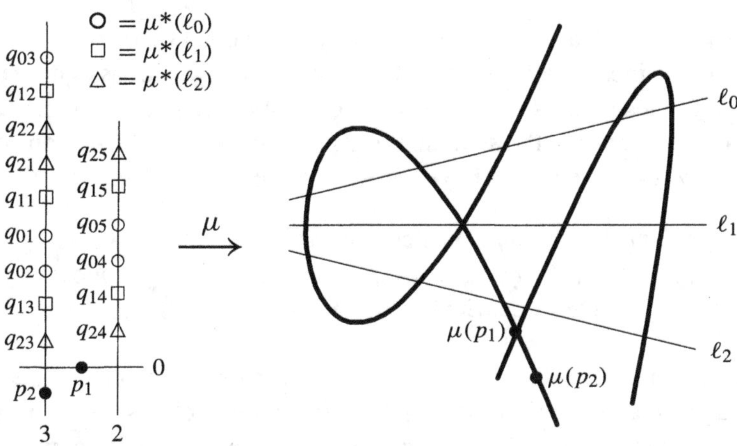

Each of the three divisors $D_j = \mu^*\ell_j$ is distributed with 3 points on the twig of degree 3, no point on the twig of degree zero, and 2 points on the twig of degree 2. Note that nothing prevents the "node" of the cubic part from falling on top of the line ℓ_1. What matters is that the inverse image in C consists of distinct and nonspecial points.

Note that these open sets do in fact cover $\overline{M}_{0,n}(\mathbb{P}^2, d)$; that is, for every map $\mu : C \to \mathbb{P}^2$ there exists a choice of three lines such that μ belongs to the corresponding open set. The existence of such three lines is evident if the restriction of μ to each twig is birational or constant: do not take any tangent line or any line through the image of any special point. If the map is a multiple cover, you must also avoid lines passing through the images of the ramification points.

2.4.4 The companion stable m-pointed curve. To each map $\mu : C \to \mathbb{P}^2$ satisfying the transversality condition (2.4.2) we associate an m-pointed rational curve $\widetilde{C}$, with $m = n + 3d$. The curve is simply the source curve C, and the m marked points are the n original marked points, supplemented by further $3d$ marked points obtained as the inverse images of the three lines. These extra marked points will be denoted by $q_{j1}, \dots, q_{jd}$, $0 \leq j \leq 2$. Now we claim that *the constructed curve $\widetilde{C}$ is stable as an m-pointed curve if and only if $\mu : C \to \mathbb{P}^2$ is stable as a map.*

Indeed, suppose μ is Kontsevich stable. Then by definition, any twig of degree zero is already stable as a pointed curve. On any twig of degree $d_A > 0$ there are $3d_A \geq 3$ new marked points on $\widetilde{C}$, distinct from the special points, and this ensures the stability of such a twig as a pointed curve. Conversely, if μ were not Kontsevich stable, there would be a twig of degree zero with fewer than three special points, leaving C unstable as a pointed curve. Since this twig is of degree zero, there would be no further marked points on it as a twig of $\widetilde{C}$, which would therefore be unstable as well.

2.4.5 Remark. Note that there is an ambiguity in the construction of $\widetilde{C}$: while each divisor D_j determines a well-defined set of d points, the way of assigning the marks $q_{j1}, \ldots, q_{jd}$ to these points is not given canonically. Permuting the marks (with j fixed), the divisor remains the same, but we get potentially $d!\,d!\,d!$ nonisomorphic m-pointed curves $\widetilde{C}$. We will come back to this question below.

2.4.6 The open set $B \subset \overline{M}_{0,m}$. Which are the stable curves with marked points $p_1, \ldots, p_n,\ q_{01}, \ldots, q_{0,d},\ q_{11}, \ldots, q_{1,d},\ q_{21}, \ldots, q_{2,d}$ that appear in this way? The condition is that *the three divisors defined as $D_j := \sum q_{jk}$ must be linearly equivalent.*

Indeed, we have already noted that the constructed curves enjoy this property. Conversely, given a curve $\widetilde{C}$ satisfying the requirement, choose isomorphisms between the three line bundles $\mathcal{O}(D_j)$. The divisors arise from three sections $\tilde{s}_0, \tilde{s}_1, \tilde{s}_2$ of this identified line bundle. These sections define a morphism $\tilde{\mu} : \widetilde{C} \to \mathbb{P}^2$ of degree d, since they do not vanish simultaneously. Composing with a change of coordinates $\phi \in \mathrm{Aut}(\mathbb{P}^2)$, we can assume that the three divisors are the inverse images of the three original lines. Now forget the marks q_{jk} (without stabilizing), and let $\mu : C \to \mathbb{P}^2$ be the map with only the n marked points $p_1, \ldots, p_n$. Then μ is a map that induces $\widetilde{C}$, and by the observation 2.4.4, it is then a *stable* map, since $\widetilde{C}$ is stable as an m-pointed curve.

This subset of $\overline{M}_{0,m}$ will be denoted by B. Note that B certainly contains all the irreducible m-pointed curves, since in $\mathbb{P}^1$, the equivalence class of a line bundle is determined by its degree. A necessary and sufficient condition for an m-pointed curve $(\widetilde{C}, (p_i), (q_{jk})) \in \overline{M}_{0,m}$ to lie in B is that it be *balanced* in the following sense: the number of points of the divisor D_j belonging to each twig of $\widetilde{C}$ *is independent of j*. In other words, the three divisors D_j are equally distributed on the twigs, with the same degree. The complement of B is the union of boundary divisors $D(A|A')$ such that A intersects some D_j in fewer points than some other $D_{j'}$. In this way we see that B is a nonempty *open* subset of $\overline{M}_{0,m}$.

2.4.7 Remark. Many nonisomorphic stable maps $\mu : C \to \mathbb{P}^2$ can induce the same m-pointed curve $(\widetilde{C}, (p_i), (q_{jk})) \in \overline{M}_{0,m}$. Indeed, consider the map

$$C \xrightarrow{\mu} \mathbb{P}^2 \xrightarrow{\phi} \mathbb{P}^2$$

where ϕ is an automorphism of $\mathbb{P}^2$ that leaves the three lines invariant. In homogenous coordinates we have $\phi([x_0 : x_1 : x_2]) = [\lambda_0 x_0 : \lambda_1 x_1 : \lambda_2 x_2]$, multiplication by an invertible diagonal matrix. Clearly we can assume $\lambda_0 = 1$. The m-pointed curve associated to the composition $\phi \circ \mu$ is equal to the curve $\widetilde{C}$ associated to

μ. This shows that there is a $\mathbb{C}^* \times \mathbb{C}^*$ of nonisomorphic maps inducing the same m-pointed curve.

From the viewpoint of the curve $\widetilde{C}$, we note the same phenomenon. At the step where we construct the map $\tilde{\mu}$, we need to specify isomorphisms among the three line bundles $\mathcal{O}(D_j)$. In other words, the map $[s_0 : s_1 : s_2]$ defined by the three sections s_0, s_1, s_2 is as good as the map $[\lambda_0 s_0 : \lambda_1 s_1 : \lambda_2 s_2]$ given by any other choice of weights $\lambda_j \in \mathbb{C}^*$.

The possible choices of weights form a $(\mathbb{C}^* \times \mathbb{C}^*)$-bundle over B. Denote by Y the total space of this bundle.

2.4.8 The quotient Y/G. Set $G = \mathfrak{S}_d \times \mathfrak{S}_d \times \mathfrak{S}_d$, the product of three copies of the symmetric group in d letters. The group G acts on Y permuting $q_{j1}, \ldots, q_{jd}$ (for each fixed j). We already saw in 2.4.5 that these permutations do not alter the section s_j, but they may alter the m-pointed curve $\widetilde{C}$. Identifying $\widetilde{C}$ with $g \cdot \widetilde{C}$ for $g \in G$, that is, passing to the quotient Y/G, we get a bijection with the open subset of $\overline{M}_{0,n}(\mathbb{P}^2, d)$ described in 2.4.2.

Check once again the dimension count:

$$\dim Y = \underbrace{2}_{\text{for the fiber } \mathbb{C}^{*2}} + \underbrace{m-3}_{\text{dimension of the base}} = \underbrace{n+3d-1}_{\text{dimension of } \overline{M}_{0,n}(\mathbb{P}^2,d)} .$$

2.4.9 Smoothness of $\overline{M}^*_{0,n}(\mathbb{P}^2, d)$. Let us argue now why the space $\overline{M}^*_{0,n}(\mathbb{P}^2, d)$ (of maps without automorphisms) is smooth. We know that when a finite group acts on a smooth variety, for each point where the action is free (i.e., the cardinality of the orbit equals the order of the group), the image point down in the quotient is also smooth (cf. Mumford [63], §7.) In the case of the action of G on Y described above, to say that the action is not free is to say that some permutation of the marks q_{jk} is induced by an automorphism of the the curve C (and fixing the n marks p_i).

Now, an automorphism of C that fixes the marked points p_i and permutes the marked points q_{jk} (for each j) is also compatible with any of the n-pointed maps $\mu : C \to \mathbb{P}^2$ corresponding to the points in Y lying over C. And conversely, given an automorphism of the map $\mu : C \to \mathbb{P}^2$, then in particular it is an automorphism of C that fixes the marked points p_i, and since it is compatible with the map μ, its effect on the new marked points q_{jk} is nothing but permutation (for each j fixed).

2.5 Evaluation maps

For each mark p_i there is a natural map,

$$\begin{array}{rcl} \nu_i : \overline{M}_{0,n}(\mathbb{P}^r, d) & \longrightarrow & \mathbb{P}^r \\ (C; p_1, \ldots, p_n; \mu) & \longmapsto & \mu(p_i) \end{array}$$

called the *evaluation map*, which is in fact a morphism.

2.5.1 Lemma. *The evaluation maps are flat.*

Proof. Each evaluation map is clearly invariant under the action of $\mathrm{Aut}(\mathbb{P}^r)$. By generic flatness (cf. [1]), and since the action on $\mathbb{P}^r$ is transitive, the map is then flat. □

Despite their apparent banality, the evaluation maps play a decisive role: they allow us to relate the geometry of $\mathbb{P}^r$ to the geometry of $\overline{M}_{0,n}(\mathbb{P}^r, d)$.

2.5.2 Example. If $H \subset \mathbb{P}^r$ is a hyperplane, then for each i the inverse image $\nu_i^{-1}(H)$ is a divisor in $\overline{M}_{0,n}(\mathbb{P}^r, d)$, consisting of all maps whose ith marked point is mapped into H.

If $Q \in \mathbb{P}^2$ is a point, then the inverse image

$$\nu_i^{-1}(Q) = \{\mu \mid \mu(p_i) = Q\}$$

is of codimension 2 in $\overline{M}_{0,n}(\mathbb{P}^2, d)$.

2.5.3 Observation. Taking the product of all the evaluation maps we get a "total evaluation map"

$$\begin{aligned} \underline{\nu} : \overline{M}_{0,n}(\mathbb{P}^r, d) &\longrightarrow \mathbb{P}^r \times \cdots \times \mathbb{P}^r \\ \mu &\longmapsto \big(\mu(p_1), \ldots, \mu(p_n)\big). \end{aligned}$$

We are going to use this viewpoint in Chapter 4. It should be noted that this map is not flat, as the example below shows.

2.5.4 Example. Consider $\overline{M}_{0,2}(\mathbb{P}^2, d)$ and let $Q \in \mathbb{P}^2$. Now the inverse image

$$\underline{\nu}^{-1}(Q, Q) = \{\mu \mid \mu(p_1) = \mu(p_2) = Q\}$$

ought to be of codimension 4 if $\underline{\nu}$ were flat, but in fact it contains a component of codimension 3, which we now describe. The general map in this locus has two twigs: one twig of degree d whose image passes through Q, and another twig of degree 0 that carries the two marked points and is attached to the first twig at the inverse image of Q (thus both marked points map to Q as required). To see that this locus has codimension 3 we will need the description of the boundary given in 2.7 below, but intuitively the count is this: one codimension because there is 1 node (cf. also 1.5.1), and two codimensions to make that node map to Q.

2.6 Forgetful maps

2.6.1 Forgetful maps. As in the case of stable curves, we may also define for each choice of sets of marks $B \subset A$ a forgetful map $\overline{M}_{0,A}(\mathbb{P}^r, d) \to \overline{M}_{0,B}(\mathbb{P}^r, d)$ that omits the marks in the complement $A \smallsetminus B$. Each forgetful map factors through forgetful maps that omit just one mark at a time,

$$\varepsilon : \overline{M}_{0,n+1}(\mathbb{P}^r, d) \to \overline{M}_{0,n}(\mathbb{P}^r, d).$$

Clearly it does not matter in which order the marks are forgotten.

The way a forgetful map affects a map with reducible source curve is similar to the case of Deligne–Mumford–Knudsen curves: twigs that become unstable by the absence of the suppressed mark must be contracted. Note that this can happen only for twigs of degree zero, since twigs of positive degree are always Kontsevich stable regardless of their marks. For this reason, the new map $\varepsilon(\mu)$ is certainly well defined: since μ was already constant on the twig in question, the image of the map does not change.

Here are some figures forgetting the marked point p_{n+1}:

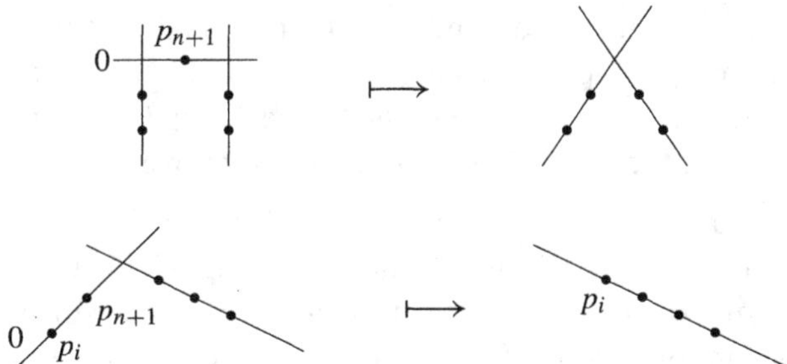

There is a close relationship between a moduli point $[\mu] \in \overline{M}_{0,n}(\mathbb{P}^r, d)$ (represented by a map $\mu : C \to \mathbb{P}^r$) and the restriction of ν_{n+1} to the fiber $F_\mu := \varepsilon^{-1}([\mu])$ of the forgetful map. In the case where μ is automorphism-free, we shall see that the relationship is a canonical identification of $\nu_{n+1\,|F_\mu}$ with μ. In the presence of automorphisms, the situation is subtler.

2.6.2 Universal family over $\overline{M}^*_{0,n}(\mathbb{P}^r, d)$. For simplicity, let us consider the case without marked points, $n = 0$. Consider first a map $\mu : C \to \mathbb{P}^r$ with smooth source curve, mapping birationally onto its image. Certainly all the 1-pointed maps belonging to the fiber F_μ of ε have these two properties. It is clear that for each choice of the marked point $p_1 \in C$ we have a 1-pointed map, and that the 1-pointed maps produced in this way are nonisomorphic. Therefore there is a natural bijection between the points of C and the points of the fiber F_μ: to each point $q \in C$ associate the 1-pointed map $\mu_q : C \to \mathbb{P}^r$ obtained from μ by setting $p_1 := q$.

More is true: the evaluation map $\nu_1 : \overline{M}_{0,1}(\mathbb{P}^r, d) \to \mathbb{P}^r$ restricted to F_μ can be identified with the map μ itself. Indeed, let $q \in C$ be any point. The corresponding point in F_μ is represented by the 1-pointed map $\mu_q : C \to \mathbb{P}^r$. Now evaluate ν_1 at it: $\nu_1([\mu_q]) = \mu_q(p_1)$. But μ_q was defined precisely to be equal to μ, except for the fact that its domain has acquired the mark $p_1 = q$. Hence, $\mu_q(p_1) = \mu(q)$ as asserted.

Let us construct formally the isomorphism $C \xrightarrow{\sim} F_\mu$. In order to have a map from C into $\overline{M}_{0,1}(\mathbb{P}^r, d)$ it is enough to exhibit a family of 1-pointed maps with base C (then the classifying map of the family gives what we want). We simply take

$$\begin{array}{ccc} C \times C & \xrightarrow{\overline{\mu}} & \mathbb{P}^r \\ \delta \uparrow \downarrow \pi & & \\ C & & \end{array}$$

where the map π is first projection and δ is the diagonal section and $\overline{\mu}(q, q') = \mu(q')$. This is a family of 1-pointed stable maps. Hence there exists a morphism $C \to \overline{M}_{0,1}(\mathbb{P}^r, d)$, whose image is precisely the fiber F_μ. It is clear that this morphism gives the set-theoretic bijection described above.

Let us proceed to a slightly less simple case. Let $\mu : C \to \mathbb{P}^r$ be a map with a reducible domain, still assumed birational to its image. This time, if q is a node, simply declaring $p_1 := q$ does not produce a stable map, since marked points are required to be smooth. Nevertheless, we do know that there is a well-defined stabilization (cf. 1.3.1). Thus the set-theoretic bijection $C \leftrightarrow F_\mu$ still holds even when C is singular. Note that the new twig introduced by the stabilization is of degree zero: it gets contracted by the map μ_q. Having said that, it is easy to see that the identification of ν_1 restricted to F_μ with μ also stays valid. Finally, the morphism $C \xrightarrow{\sim} F_\mu$ is built as in the smooth case, except that the diagonal section no longer avoids the singularities of the fibers, so that some blowups are needed to achieve the family of stable 1-pointed maps with base C.

The case in which there are marked points leads to similar problems: the diagonal section intersects the constant sections corresponding to the marked points, and it is necessary to blow up these intersection points. These considerations show that, when restricted to the open set $\overline{M}^*_{0,n}(\mathbb{P}^r, d)$ of automorphism-free maps, our forgetful map ε plays the role of a tautological family. In fact, we are dealing with the universal family, recalling the assertion of Theorem 2.3.3 to the effect that $\overline{M}^*_{0,n}(\mathbb{P}^r, d)$ is a fine moduli space.

2.6.3 The fibers of ε in the presence of automorphisms. For simplicity, we take our favorite example 2.1.13 of a map with automorphisms:

$$\begin{array}{ccc} C := \mathbb{P}^1 & \xrightarrow{\mu} & \mathbb{P}^2 \\ [x:y] & \longmapsto & [x^2 : y^2 : 0]. \end{array}$$

Following the same procedure as the one described on the previous page, we construct a map $\rho : C \to \overline{M}_{0,1}(\mathbb{P}^2, 2)$ by looking at $C \times C$ with the diagonal section. However in this case, *the map ρ is not injective.* The reason is the presence of nontrivial automorphisms. Indeed, consider the automorphism $\phi([x : y]) = [-x : y]$ which respects μ. Pick a point $q \in C$ distinct from the two ramification points of μ, and consider the corresponding 1-pointed map μ_q. Compare with the map corresponding to the point $\phi(q) \in C$. These are two distinct 1-pointed maps, but *they are isomorphic*, since the $\mathbb{P}^1$-automorphism ϕ transforms one map into the other. Therefore q and $\phi(q)$ have the same image in $\overline{M}_{0,1}(\mathbb{P}^2, 2)$, that is, the map $\rho : C \to \overline{M}_{0,1}(\mathbb{P}^2, 2)$ is a $2 : 1$ cover.

Now let us compare μ with the evaluation map ν_1 restricted to the fiber F_μ. They are related by the following factorization:

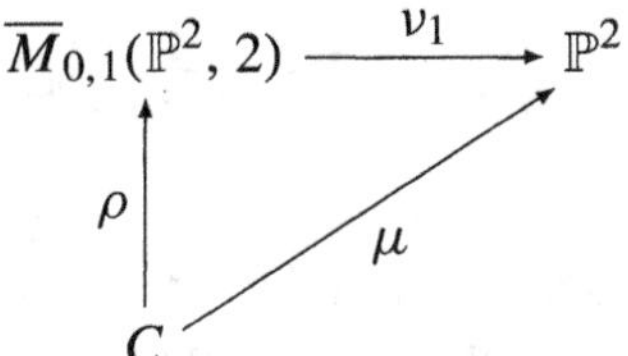

Indeed, the image $\rho(q)$ of a point $q \in C$ is the 1-pointed map $\mu_q : C \to \mathbb{P}^2$, which is simply the original map μ equipped with the marked point $p_1 := q$. Now we must evaluate this map at p_1. But $p_1 = q$, so the result is just $\mu(q)$. Note in particular that ν_1 restricted to F_μ is bijective onto its image. If ν_1 were to be the universal map (as in the automorphism-free case), its restriction to that fiber would be a double cover. But it is in fact a bijective map from a double curve (F_μ) onto the image. In fact, the scheme-theoretic fiber of ε over the point $[\mu]$ is nonreduced. In particular, $\varepsilon : \overline{M}_{0,1}(\mathbb{P}^2, 2) \longrightarrow \overline{M}_{0,0}(\mathbb{P}^2, 2)$ is not even a family of stable maps.

We shall get the opportunity to encounter this phenomenon again in Section 2.9.

2.6.4 Incidences. Inside $\overline{M}_{0,n+1}(\mathbb{P}^r, d)$, consider the locus $\nu_{n+1}^{-1}(H^k)$ of all the maps μ such that $\mu(p_{n+1}) \in H^k$, where $H^k \subset \mathbb{P}^r$ is a linear subspace of codimension $k \geq 2$. Forgetting the mark p_{n+1} we get, in a space with one mark less, the locus of maps that are just incident to H^k, without mention of marks. To be precise,

$$\operatorname{inc}(H^k) := \varepsilon\big(\nu_{n+1}^{-1}(H^k)\big)$$

is the subvariety in $\overline{M}_{0,n}(\mathbb{P}^r, d)$ of codimension $k - 1$ consisting of all the maps incident to H^k. In particular, $\operatorname{inc}(H^2)$ is an important divisor.

2.6.5 Example. Since maps of degree one have no automorphisms, the forgetful map $\varepsilon : \overline{M}_{0,1}(\mathbb{P}^r, 1) \to \overline{M}_{0,0}(\mathbb{P}^r, 1)$ is a universal family. We have already seen that the base $\overline{M}_{0,0}(\mathbb{P}^r, 1)$ can be identified with the Grassmannian $\mathrm{Gr}(1, \mathbb{P}^r)$, and likewise, $\overline{M}_{0,1}(\mathbb{P}^r, 1)$ is precisely the universal line. If $H^k \subset \mathbb{P}^r$ is a linear subspace of codimension k, the inverse image $\nu_1^{-1}(H^k) \subset \overline{M}_{0,1}(\mathbb{P}^r, 1)$ is the total space of the family of lines incident to H^k, and $\mathrm{inc}(H^k) = \varepsilon\big(\nu_1^{-1}(H^k)\big) \subset \overline{M}_{0,0}(\mathbb{P}^r, 1)$ is therefore identified with the Schubert variety $\Sigma_0(H^k) \subset \mathrm{Gr}(1, \mathbb{P}^r)$ (see Harris [42]).

2.6.6 Forgetting the map to $\mathbb{P}^r$. For $n \geq 3$, there is also a forgetful map

$$\overline{M}_{0,n}(\mathbb{P}^r, d) \to \overline{M}_{0,n}$$

consisting in forgetting the data of the map to $\mathbb{P}^r$ and stabilizing, contracting twigs that become unstable. This map can be constructed locally, using the open cover of 2.4.8 of type Y/G. Since Y/G is a fibration over $\overline{M}_{0,m}$, there is a usual forgetful map (of Deligne–Mumford–Knudsen spaces) to $\overline{M}_{0,n}$, and since the action of G consists in permuting the marks we are forgetting, this morphism is obviously G-invariant, inducing $Y/G \to \overline{M}_{0,n}$.

2.6.7 Lemma. *For $n \geq 4$, the forgetful map $\overline{M}_{0,n}(\mathbb{P}^r, d) \to \overline{M}_{0,4}$ is a flat morphism.*

Proof. Flatness of a reduced and irreducible variety over a nonsingular curve such as $\overline{M}_{0,4} = \mathbb{P}^1$ is rather easy: it suffices that the map be surjective (dominating is enough), cf. [44, p. 257]. □

2.6.8 Remark. More generally, for $n \geq 3$, the forgetful map

$$\eta : \overline{M}_{0,n}(\mathbb{P}^r, d) \to \overline{M}_{0,n}$$

is a flat morphism.

Indeed, as we have recalled in 2.6.6, an open neighborhood of $\overline{M}_{0,n}(\mathbb{P}^r, d)$ can be taken of the form $V = Y/G$. By construction, it is clear that $\eta_{|V}$ fits into the commutative diagram

$$\begin{array}{ccc} Y & \longrightarrow & \overline{M}_{0,m} \\ \downarrow & & \downarrow \\ V & \xrightarrow{\eta_{|V}} & \overline{M}_{0,n} \end{array}$$

where the right-hand vertical arrow is a forgetful map for Deligne–Mumford–Knudsen spaces, known to be flat. Thus, we have reduced our claim to the following statement. *Let Y be a variety with an action of a finite group G. Let $\varphi : Y/G \to Z$ be a morphism such that the composite $Y \to Y/G \to Z$ is a flat morphism. Then φ is flat.* Indeed, this translates into homomorphisms of coordinate rings,

$R \to A \to B$, where B is G-invariant, $A = B^G$ is the ring of invariants, and B is flat over R. Now invoke the A-homomorphism $\rho : B \to A$ defined by "averaging," $\rho(b) = \frac{1}{|G|}\sum_g gb$. This is a "retraction" for the inclusion map $A \to B$, i.e., $\rho(b) = b$ for $b \in A$. It follows that A can be identified with a direct summand of B (as an A-module) and therefore is R-flat.

2.7 The boundary

The boundary of $\overline{M}_{0,n}(\mathbb{P}^r, d)$ is formed by maps whose domains are reducible curves. The description of the boundary is very similar to the one we have given for the boundary of $\overline{M}_{0,n}$. It boils down to the combinatorics of distribution of marked points and degrees.

Definition. A d-weighted partition of a set $S := \{p_1, \ldots, p_n\}$ consists of a partition $A \cup B = S$ together with a partition $d_A + d_B = d$ into nonnegative integers.

2.7.1 Boundary divisors. For each d-weighted partition

$$A \cup B = S, d_A + d_B = d \text{ (where } \sharp A \geq 2 \text{ if } d_A = 0, \text{ and } \sharp B \geq 2 \text{ if } d_B = 0)$$

there exists an irreducible divisor, denoted $D(A, B; d_A, d_B)$, called a *boundary divisor.* A general point on this divisor represents a map μ whose domain is a tree with two twigs, $C = C_A \cup C_B$, with the points of A in C_A and those of B in C_B, such that the restriction of μ to C_A is a map of degree d_A and the restriction of μ to C_B is of degree d_B. We shall indicate it by a picture as follows:

We have the following counterpart to 1.5.8.

2.7.2 Proposition. *The union of the boundary divisors in $\overline{M}_{0,n}(\mathbb{P}^r, d)$ is a divisor with normal crossings, up to a finite quotient. (That is, their irreducible components meet transversally up to a finite quotient.)* □

Exercise 18 on page 90 gives a formula for the number of boundary divisors. For instance, $\overline{M}_{0,5}(\mathbb{P}^2, 2)$ has 42 boundary divisors; $\overline{M}_{0,8}(\mathbb{P}^2, 3)$ has 503; and $\overline{M}_{0,11}(\mathbb{P}^2, 4)$ has 5108.

2.7.3 Recursive structure. From the combinatorial description of the boundary we get a natural gluing morphism

$$\overline{M}_{0,A\cup\{x\}}(\mathbb{P}^r, d_A) \times_{\mathbb{P}^r} \overline{M}_{0,B\cup\{x\}}(\mathbb{P}^r, d_B) \longrightarrow D(A, B; d_A, d_B), \qquad (2.7.3.1)$$

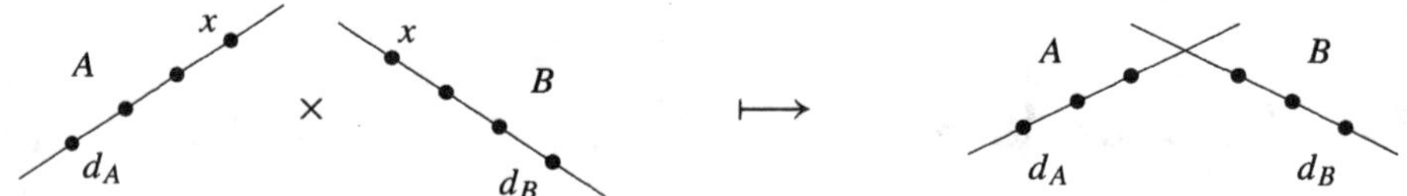

The fiber product over $\mathbb{P}^r$ is taken via the evaluation maps at the marked point x,

$$\begin{aligned} \nu_{x_A} : \overline{M}_{0,A\cup\{x\}}(\mathbb{P}^r, d_A) &\longrightarrow \mathbb{P}^r \\ \nu_{x_B} : \overline{M}_{0,B\cup\{x\}}(\mathbb{P}^r, d_B) &\longrightarrow \mathbb{P}^r. \end{aligned}$$

This is but expressing in fancy words the requirement that the marked point indicated by x must have the same image in $\mathbb{P}^r$ under both maps in order for the gluing to make sense.

The morphism 2.7.3.1 is in fact an isomorphism, with few exceptions. Only in some very special cases, can the presence of symmetries render it noninjective. In one case, $A = B = \emptyset$, $d_A = d_B$, the situation is so symmetric that the map is in fact generically 2-1. See FP-NOTES, Lemma 12, and [52], Lemma 2.2 for a precise statement.

The fiber product can be seen as a subvariety of the usual product

$$\overline{M}_{0,A\cup\{x\}}(\mathbb{P}^r, d_A) \times \overline{M}_{0,B\cup\{x\}}(\mathbb{P}^r, d_B)$$

given by the inverse image of the diagonal $\Delta \subset \mathbb{P}^r \times \mathbb{P}^r$:

$$\overline{M}_{0,A\cup\{x\}}(\mathbb{P}^r, d_A) \times_{\mathbb{P}^r} \overline{M}_{0,B\cup x}(\mathbb{P}^r, d_B) = (\nu_{x_A} \times \nu_{x_B})^{-1}(\Delta).$$

In this way, intersections with $D(A, B; d_A, d_B)$ can be computed in the spaces $\overline{M}_{0,A\cup\{x\}}(\mathbb{P}^r, d_A)$ and $\overline{M}_{0,B\cup\{x\}}(\mathbb{P}^r, d_B)$, whose dimensions are strictly smaller. This fact will be crucial in the remaining chapters (cf. 4.3.2).

2.7.4 Remark. Note that even if we had started with a space without any marks, we would be forced to consider marks in order to describe its boundary.

2.7.5 Special boundary divisors. For $n \geq 4$, consider the composition of forgetful maps $\overline{M}_{0,n}(\mathbb{P}^r, d) \to \overline{M}_{0,n} \to \overline{M}_{0,4}$, which we know is flat, cf. 2.6.7. Let $D(ij|kl)$ be the divisor in $\overline{M}_{0,n}(\mathbb{P}^r, d)$ defined as the inverse image of the divisor $(ij|kl)$ in $\overline{M}_{0,4}$. Then

$$D(ij|kl) = \sum D(A, B; d_A, d_B),$$

where the sum is taken over all d-weighted partitions of the marking set $S = \{p_1, \ldots, p_n\}$ such that $i, j \in A$ and $k, l \in B$. By a reasoning similar to that indicated in 1.5.10, all the coefficients in this sum are equal to one. Recalling that in $\overline{M}_{0,4} \simeq \mathbb{P}^1$ all three boundary divisors are equivalent, we obtain the *fundamental relation*

$$\sum_{\substack{A \cup B = S \\ i,j \in A \\ k,l \in B \\ d_A + d_B = d}} D(A, B; d_A, d_B) \equiv \sum_{\substack{A \cup B = S \\ i,k \in A \\ j,l \in B \\ d_A + d_B = d}} D(A, B; d_A, d_B) \equiv \sum_{\substack{A \cup B = S \\ i,l \in A \\ j,k \in B \\ d_A + d_B = d}} D(A, B; d_A, d_B), \tag{2.7.5.1}$$

the consequences of which will be explored in the remaining chapters.

2.8 Easy properties and examples

The first couple of results are mere exercises about the evaluation maps. Next we take a closer look at the simpler spaces, viz., $d = 0$ and $d = 1$.

2.8.1 Compatibility between recursive structure and evaluation maps. Consider a boundary divisor $D = D(A, d_A; B, d_B)$ for some partition $A \cup B = S$ and suppose the mark p_i is in A. Then we have the following commutative diagram:

$$\begin{array}{ccc} \overline{M}_{0,A\cup\{x\}}(\mathbb{P}^r, d_A) \times_{\mathbb{P}^r} \overline{M}_{0,B\cup x}(\mathbb{P}^r, d_B) & \longrightarrow & D \subset \overline{M}_{0,n}(\mathbb{P}^r, d) \\ \downarrow & & \downarrow \nu_i \\ \overline{M}_{0,A\cup\{x\}}(\mathbb{P}^r, d_A) & \xrightarrow[\nu_i]{} & \mathbb{P}^r \end{array}$$

where the left-hand arrow is the projection, and the bottom arrow is the evaluation of the mark p_i in A.

2.8.2 Lemma. *Let $\Gamma \subset \mathbb{P}^r$ be a subvariety. Then its inverse image $\nu_i^{-1}\Gamma \subset \overline{M}_{0,n}(\mathbb{P}^r, d)$ has proper intersection with each boundary divisor D. That is, if Γ is of codimension k then the intersection $\nu_i^{-1}\Gamma \cap D$ is of codimension $k + 1$ in $\overline{M}_{0,n}(\mathbb{P}^r, d)$.*

Proof. Consider a boundary divisor $D = D(A, B; d_A, d_B)$ where $p_i \in A$. Using the finite gluing morphism

$$\overline{M}_{0,A\cup\{x\}}(\mathbb{P}^r, d_A) \times_{\mathbb{P}^r} \overline{M}_{0,B\cup\{x\}}(\mathbb{P}^r, d_B) \longrightarrow D(A, B; d_A, d_B)$$

and the compatibility with evaluation maps, we see that the intersection $D \cap \nu_i^{-1}\Gamma$ is the image of $\nu_{A_i}^{-1}\Gamma \times_{\mathbb{P}^r} \overline{M}_{0,B\cup\{x\}}(\mathbb{P}^r, d_B)$, where ν_{A_i} is the map of evaluation at the mark $p_i \in A$ for the space $\overline{M}_{0,A\cup\{x\}}(\mathbb{P}^r, d_A)$. Flatness of this map ensures that $\nu_{A_i}^{-1}\Gamma$ is of codimension k in $\overline{M}_{0,A\cup\{x\}}(\mathbb{P}^r, d_A)$; hence $D \cap \nu_i^{-1}\Gamma$ is of codimension $k+1$ as asserted. □

2.8.3 Lemma. *The fibers of* ν_i *are irreducible.*

Proof. We reduce first to the case of many marked points. In the diagram

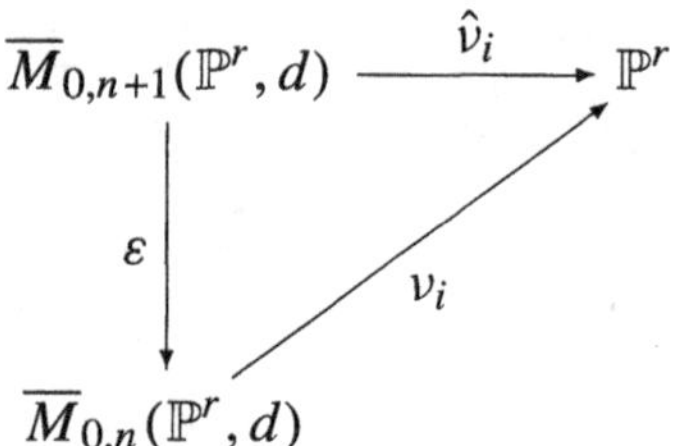

ε is the forgetful map of p_{n+1}, whereas $\hat{\nu}_i$ and ν_i are the maps of evaluation at p_i for the respective spaces. Clearly the diagram commutes, and so $\hat{\nu}_i^{-1}(\Gamma) = \varepsilon^{-1}\nu_i^{-1}(\Gamma)$. Now if $\nu_i^{-1}(\Gamma)$ were reducible, then $\varepsilon^{-1}\nu_i^{-1}(\Gamma)$ would also be so. Hence, the validity of the assertion for the space with $n+1$ marks implies the result for the case of n marks.

Therefore we may assume $n \geq 3$. The fiber is a subscheme of $\overline{M}_{0,n}(\mathbb{P}^r, d)$ of codimension r, and knowing from the previous lemma that it intersects the boundary properly, it is sufficient to show irreducibility of its inverse image in the open subset $M_{0,n}(\mathbb{P}^r, d)$ formed by maps with domain $\mathbb{P}^1$. We now use the description

$$M_{0,n}(\mathbb{P}^r, d) \simeq M_{0,n} \times W(r, d)$$

given in 2.1.16. In view of the transitive action of $\mathrm{Aut}(\mathbb{P}^r)$ on $\mathbb{P}^r$, it is enough to establish the irreducibility of the fiber over one point, say $0 = [0, \ldots, 0, 1] \in \mathbb{P}^r$. We may also assume that $p_i \in \mathbb{P}^1$ is the point $[0 : 1]$. Now the condition that $\mu([0 : 1]) = 0$ means that the first r binary forms defining μ vanish at $[0 : 1]$. This certainly amounts to r linearly independent conditions. Hence the fiber is an open subset of a linear subspace in $M_{0,n} \times W(r, d)$ and so is irreducible. □

2.8.4 Corollary. *For any irreducible subvariety* $\Gamma \subset \mathbb{P}^r$*, its inverse image under evaluation is irreducible.*

Proof. Since the fibers of $\nu_i^{-1}(\Gamma) \to \Gamma$ are irreducible and of constant dimension, the irreducibility of $\nu_i^{-1}(\Gamma)$ follows at once. □

2.8.5 Degree 0. Even though our main interest is the nonconstant maps $\mathbb{P}^1 \to \mathbb{P}^r$ (which yield honest curves), it is necessary to understand the degenerate behavior of the case $d = 0$. A stable map of degree 0 sends the whole source curve onto a single point. Since its source must be a stable pointed curve, we certainly have $n \geq 3$. This leads us to consider the two natural morphisms

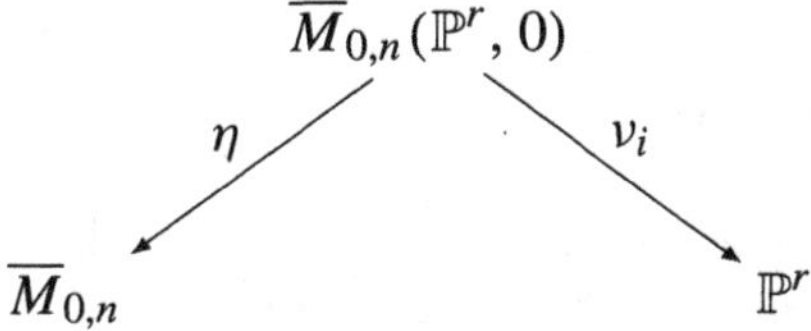

where η is the forgetful map (cf. 2.6.6), whereas ν_i is any of the evaluation maps; they coincide. In this case, the map η does not involve contraction and you may easily verify that the product of these two maps gives in fact an isomorphism

$$\overline{M}_{0,n}(\mathbb{P}^r, 0) \xrightarrow{\sim} \overline{M}_{0,n} \times \mathbb{P}^r.$$

Note in particular that for $r = 0$, we have $\mathbb{P}^0 = \operatorname{Spec} \mathbb{C}$, and so the spaces of Kontsevich include all the Deligne–Mumford–Knudsen spaces studied in Chapter 1.

2.8.6 Degree 1 (no marks or a single mark). We shall explore the property of coarse moduli space, in order to give a more formal identification of the space $\overline{M}_{0,0}(\mathbb{P}^r, 1)$ with the Grassmannian of lines in the projective space $\mathbb{P}^r$. Consider the universal family of lines

$$\begin{array}{ccc} U & \subset & \operatorname{Gr}(1, \mathbb{P}^r) \times \mathbb{P}^r \\ {\scriptstyle \pi}\downarrow & & \downarrow \\ \operatorname{Gr}(1, \mathbb{P}^r) & & \mathbb{P}^r \end{array}$$

Since π is a $\mathbb{P}^1$-bundle, we are clearly dealing with a flat family of smooth rational curves. Putting it together with the map $U \to \mathbb{P}^r$ we get in this way a family of stable maps of degree 1, with base $\operatorname{Gr}(1, \mathbb{P}^r)$. Now the universal property of $\overline{M}_{0,0}(\mathbb{P}^r, 1)$ ensures the existence of a map $\operatorname{Gr}(1, \mathbb{P}^r) \to \overline{M}_{0,0}(\mathbb{P}^r, 1)$. This map is obviously bijective. Since $\operatorname{Gr}(1, \mathbb{P}^r)$ is smooth, and since $\overline{M}_{0,0}(\mathbb{P}^r, 1)$ is normal, we may apply Zariski's main theorem (cf. [65, ch. III, §9]). We deduce that the map is an isomorphism. A similar argument identifies $\overline{M}_{0,1}(\mathbb{P}^r, 1)$ with the universal line U.

2.8.7 Degree one (two marks). *The space $\overline{M}_{0,2}(\mathbb{P}^r, 1)$ is naturally isomorphic to $\mathbb{P}^r \times \mathbb{P}^r$ blown up along the diagonal.*

First, we note that the blowup is easily identified with the fiber product $U \times_{\mathbb{G}} U$, where $U \to \mathbb{G} = \operatorname{Gr}(1, \mathbb{P}^r)$ is the universal line from the previous example. The two evaluation maps yield a morphism $\nu : \overline{M}_{0,2}(\mathbb{P}^r, 1) \to U \times_{\mathbb{G}} U$ that associates to each $\mu : (C, p_1, p_2) \to \mathbb{P}^r$ the pair of points $\mu(p_1), \mu(p_2)$ on the image line of μ. Note that the boundary is mapped bijectively onto the diagonal. Again by Zariski's main theorem, the morphism ν is an isomorphism.

2.9 Complete conics

We close this chapter with a somewhat detailed analysis of the space $\overline{M}_{0,0}(\mathbb{P}^2, 2)$. Our goal is to verify a result that belongs to folklore: this space is isomorphic to the variety of complete conics. (Some background on complete conics is provided in the exercises, page 87.) The example is a good occasion to practice with some of the notions we have introduced so far. A word of caution: the amount of technicalities discussed in this section is disproportional vis-à-vis the rest of the text, but neither the result nor the techniques will be required elsewhere.

2.9.1 Divisors of $\overline{M}_{0,0}(\mathbb{P}^2, 2)$. There is but one boundary divisor, which we shall denote by D. This divisor is formed by maps with domain consisting of two twigs, both of degree 1. Clearly the general element of D maps onto two distinct lines, and therefore is bijective onto its image.

Let us denote by R the locus of maps that are *not* bijective. A general element $\mu \in R$ is a double cover. R is of codimension one; this follows readily from Lemma 2.1.4, at least off the boundary. Note that R can also be characterized as the locus of maps that admit automorphisms.

The intersection $\Sigma := D \cap R$ of the two divisors described above is the locus formed by the maps with two twigs with the same image line.

2.9.2 $\overline{M}_{0,0}(\mathbb{P}^2, 2)$ is smooth. Recall from 2.4 (with the notation introduced there) that $\overline{M}_{0,0}(\mathbb{P}^2, 2)$ is locally the quotient of a smooth variety Y by the action of the finite group $G = \mathfrak{S}_2 \times \mathfrak{S}_2 \times \mathfrak{S}_2$. This action is not free precisely at the points of the subvariety $\widetilde{R} \subset Y$ that covers R. The orbit of each point of $\widetilde{R}$ has cardinality 4. More precisely, for a curve with 6 marked points that lies over a point of R, the stabilizer is the "diagonal" subgroup $H \subset G$, which is of order 2, i.e., the group that consists of the identity together with a *simultaneous* switch of the marked points,

$$q_{01} \leftrightarrow q_{02}, \quad q_{11} \leftrightarrow q_{12}, \quad q_{21} \leftrightarrow q_{22}.$$

We may take the quotient of Y by the action of G in two steps: first by the action of $H \simeq \mathbb{Z}_2$ and then by the action of G/H on $\overline{Y} := Y/H$, which now is free.

Using analytic coordinates, the action of the generator $h \in H$ in a neighborhood of a point of $\widetilde{R}$ can be written as

$$\begin{array}{ccc} \mathbb{C}[[x_1, \dots, x_5]] & \xrightarrow{h} & \mathbb{C}[[x_1, \dots, x_5]] \\ f(x_1, \dots, x_5) & \longmapsto & f(-x_1, \dots, x_5). \end{array}$$

Indeed, one knows that any action of $\mathbb{Z}_2$ (or more generally, of any finite group) on $\mathbb{C}[[x_1, \dots, x_n]]$ is linearizable. The ring of invariants is $\mathbb{C}[[x_1^2, x_2, \dots, x_5]]$. Since this is a regular ring, the claim of smoothness of the quotient is proved.

2.9.3 Complete conics. We give a brief survey on complete conics; some more details are given in the exercises, page 87. For each nondegenerate conic $C \subset \mathbb{P}^2$, the set of its tangent lines is parametrized by another conic $\check{C} \subset \check{\mathbb{P}}^2$ in the dual plane, the so-called dual conic. The collection of pairs $(C, \check{C})$ is a subvariety of $\mathbb{P}^5 \times \check{\mathbb{P}}^5$. Its closure $\mathbb{B} \subset \mathbb{P}^5 \times \check{\mathbb{P}}^5$ is the variety of complete conics. One shows that the projection $\mathbb{B} \to \mathbb{P}^5$ is the blowup along the Veronese surface V of the double lines (see Exercise 9 on page 87). In particular, $\mathbb{B}$ is a smooth variety. The fiber of the exceptional divisor $E \subset \mathbb{B}$ over a point representing a double line $2L$ is the linear system $\simeq \mathbb{P}^2$ of degree-2 divisors on L. Such two points on a double line are called *foci*. If $2L$ is the limit of a pencil of smooth conics, the dual pencil has a line pair as limit, and the foci are dual to this line pair (cf. Exercise 11 on page 88).

2.9.4 Proposition. *The variety* $\overline{M}_{0,0}(\mathbb{P}^2, 2)$ *is naturally isomorphic to the variety of complete conics.*

2.9.5 Set-theoretic description of the bijection $\overline{M}_{0,0}(\mathbb{P}^2, 2) \leftrightarrow \mathbb{B}$. For each μ in the open subset $M^*_{0,0}(\mathbb{P}^2, 2)$ the image is always a smooth conic; every such nondegenerate conic occurs exactly once.

The other possibility, still with irreducible domain, is a double cover of a line in $\mathbb{P}^2$. In this case, the two points of ramification correspond to the foci. It follows that the open subset $M_{0,0}(\mathbb{P}^2, 2)$ is in bijection with the open subset of $\mathbb{B}$ formed by the smooth conics and double lines with distinct foci. We have thus accounted for the whole divisor $E \subset \mathbb{B}$, except for the points with coincident foci.

The unique boundary divisor $D \subset \overline{M}_{0,0}(\mathbb{P}^2, 2)$ provides for all line pairs, including the case of a repeated line, whereupon we mark a single, repeated focus. We get in this way all the configurations in E that were previously omitted.

2.9.6 Idea of the formal proof. We are required to construct a morphism $\overline{M}_{0,0}(\mathbb{P}^2, 2) \to \mathbb{B}$ corresponding to the set-theoretic description given above.

Step one: construct a morphism to $\mathbb{P}^5$.

Step two: verify that the inverse image of the Veronese V is precisely the Cartier divisor R. It then follows that the map factors through the blowup of $\mathbb{P}^5$ along V, which is just $\mathbb{B}$. One more application of Zariski's main theorem ensures us that the bijection is an isomorphism.

Let us check some details.

2.9.7 Construction of the natural map $\overline{M}_{0,0}(\mathbb{P}^2, 2) \to \mathbb{P}^5$. Consider the forgetful map and the evaluation map,

$$\begin{array}{ccc} \overline{M}_{0,1}(\mathbb{P}^2, 2) & \xrightarrow{\nu_1} & \mathbb{P}^2 \\ \varepsilon \downarrow & & \\ \overline{M}_{0,0}(\mathbb{P}^2, 2). & & \end{array}$$

We get a map $\overline{M}_{0,1}(\mathbb{P}^2, 2) \to \overline{M}_{0,0}(\mathbb{P}^2, 2) \times \mathbb{P}^2$. Its image is a Cartier divisor $\mathfrak{X} \subset \overline{M}_{0,0}(\mathbb{P}^2, 2) \times \mathbb{P}^2$. Set-theoretically, it is clear that the fiber of $\mathfrak{X}$ over a point $\mu \in \overline{M}_{0,0}(\mathbb{P}^2, 2)$ is the image curve of μ (in general a nondegenerate conic). Indeed, $\mathfrak{X}$ is the total space of a *flat* family over $\overline{M}_{0,0}(\mathbb{P}^2, 2)$. This follows from (Kollár [55, 1.11]) noting that a local equation of $\mathfrak{X}$ in $\overline{M}_{0,0}(\mathbb{P}^2, 2) \times \mathbb{P}^2$ is a nonzero divisor in the fiber $\mathbb{P}^2$. Recalling that $\mathbb{P}^5$ parametrizes the universal family of conics, we obtain the morphism $\kappa : \overline{M}_{0,0}(\mathbb{P}^2, 2) \to \mathbb{P}^5$ that sends each μ to its image (be it a nondegenerate conic or a pair of lines).

2.9.8 The inverse image of the Veronese V is the Cartier divisor R. Again set-theoretically, no doubt about it. It remains to verify that the scheme inverse image presents no embedded component. There is but one place at risk for such bad behavior: Σ, the unique closed orbit of the action of $\mathrm{Aut}(\mathbb{P}^2)$ in R. A trick to detect such singularities is to draw an arc in $\overline{M}_{0,0}(\mathbb{P}^2, 2)$ that passes through Σ and compute the tangent spaces, as we proceed to explain.

2.9.9 Construction of the arc in $\overline{M}_{0,0}(\mathbb{P}^2, 2)$. The most important technique to construct an arc in a moduli space is by means of 1-parameter families. Presently, given a 1-parameter family $S \to B \times \mathbb{P}^2$ of stable maps (cf. 2.3) of degree 2 with base B, the defining property of coarse moduli space gives the classifying map $\kappa : B \to \overline{M}_{0,0}(\mathbb{P}^2, 2)$.

We start with the family of conics $bX^2 - b^2Y^2 - Z^2$, which includes as special member ($b = 0$) a double line. One checks that the dual family also presents a double line as its limit (cf. Exercise 12 on page 88).

Looking for the corresponding family of parametrizations as in Example 2.2.2, you find out that it is necessary to perform a *base change* on the family, replacing b by b^2. The family is then replaced by $b^2X^2 - b^4Y^2 - Z^2$; the conics that occur in the family are the same ones as in the original family, but now each conic appears twice, except for the special member. This one appears precisely once, due to the fact that $b \mapsto b^2$ is ramified at $b = 0$.

The good news is that now the family admits a section, given by $[b : 1 : 0]$. Just as in 2.2.2, this enables us to find the family of parametrizations

$$(b, t) \mapsto \begin{bmatrix} b(b^2 + t^2) \\ t^2 - b^2 \\ 2b^3t \end{bmatrix}.$$

This is a rational map with a base point $(b, t) = (0, 0)$. The rest of the central fiber F maps to the point $[0 : 1 : 0]$. Blowing up the base point makes the map well defined on (the strict transform of) F, but two new base points appear on the exceptional divisor E_1. Blowing up these two points resolves the map, and the two new exceptional divisors E_2 and E_3 map to the same line $(Z = 0)$.

In other words, the central fiber became a curve with four twigs: the first two (F and E_1) have degree zero and destabilize the family. Two blowdowns are required, namely, first contract F (which is a (-1)-curve), and then contract E_1 as well, which has been turned into a (-2)-curve. This last blowdown renders the total space singular, but this is irrelevant.

Now we have a family of stable maps $\widetilde{S} \to B \times \mathbb{P}^2$, and hence a map $B \to \overline{M}_{0,0}(\mathbb{P}^2, 2)$. However, each map appears twice (as was the case for the family of the images, $b^2X^2 - b^4Y^2 - Z^2$), so that $B \to \overline{M}_{0,0}(\mathbb{P}^2, 2)$ is a double cover of its image, ramified at $b = 0$. But then it factors

$$B \xrightarrow{b \mapsto b^2} B \xrightarrow{\alpha} \overline{M}_{0,0}(\mathbb{P}^2, 2),$$

where α is birational onto its image. The arc $\alpha : B \to \overline{M}_{0,0}(\mathbb{P}^2, 2)$ will be used to compute the tangent spaces. Composing with $\kappa : \overline{M}_{0,0}(\mathbb{P}^2, 2) \to \mathbb{P}^5$, we obtain exactly our original family of conics $bX^2 - b^2Y^2 - Z^2$.

Why was this base change necessary? Because there does not exist a 1-parameter family of stable maps whose corresponding family of image conics is $bX^2 - b^2Y^2 - Z^2$. However, the arc $\alpha : B \to \overline{M}_{0,0}(\mathbb{P}^2, 2)$ does exist and witnesses the following fact: in general, given a moduli space that is only coarse, a subvariety of it does not necessarily correspond to a family!

The construction given above is an example of the important technique of stable reduction, very well explained in Harris–Morrison [43, § 3C].

2.9.10 Lemma. *Let Y be a smooth variety and let $D \subset Y$ be a subscheme of codimension 1. Let B be a smooth curve. Let $\eta : B \to Y$ be a morphism such that the scheme-theoretic inverse image $\eta^{-1}D$ is a reduced point $0 \in B$. Then D is smooth at the point $p = \eta(0)$.*

Proof. Let $\mathfrak{m}_p \subset \mathcal{O}_{Y,p}$ be the ideal of the point p and let $J \subset \mathcal{O}_{Y,p}$ denote the ideal of D. Recall that the tangent space T_pY is given as $(\mathfrak{m}_p/\mathfrak{m}_p^2)^*$. The subspace T_pD is the annihilator of $(J + \mathfrak{m}_p^2)/\mathfrak{m}_p^2$. The tangent space T_pD is of codimension ≤ 1 in T_pY. If the inequality is strict, then J is contained in $\mathfrak{m}_p^2$. Since we have $\mathfrak{m}_p\mathcal{O}_{B,0} \subseteq \mathfrak{m}_0$, it follows that $J\mathcal{O}_{B,0}$ is contained in $\mathfrak{m}_0^2$, contradicting the assumption that the inverse image is reduced. □

2.9.11 Conclusion of the proof of 2.9.4. We shall apply the lemma to the arc α constructed above, in order to show that the inverse image, $\kappa^{-1}V$, of the Veronese is smooth (and in particular, a Cartier divisor in $\overline{M}_{0,0}(\mathbb{P}^2, 2)$). By the lemma, it suffices to check that $\alpha^{-1}\kappa^{-1}V$ is reduced. Since the ideal of V is generated by the 2×2 minors of the symmetric matrix associated to the conic, it is clear that its inverse image in B is reduced, given by the ideal generated by b. □

2.10 Generalizations and references

The constructions and some of the results have parallels for curves of higher genus and for arbitrary smooth projective varieties instead of $\mathbb{P}^r$, but the theory is somewhat more complicated.

2.10.1 Homogeneous varieties. Substituting $\mathbb{P}^r$ by a projective homogeneous variety, or more generally, a *convex* variety, does not require much further work. A variety X is convex when $H^1(\mathbb{P}^1, \mu^* T_X) = 0$ for all maps $\mu : \mathbb{P}^1 \to X$. The convex varieties include projective spaces, Grassmannians, flag manifolds, smooth quadrics, and products of such varieties.

Note that $A_1(X)$ (the group of dimension-1 cycles modulo rational equivalence, cf. Fulton [28], Ch. 1) may not be generated by a single class as in the case of $\mathbb{P}^r$, where $A_1(\mathbb{P}^r)$ is generated by the class of a line. So instead of just giving the degree d as in the case of $\mathbb{P}^r$, one has to specify a class $\beta \in A_1(X)$. So the spaces of stable maps are then of type $\overline{M}_{0,n}(X, \beta)$ parametrizing isomorphism classes of maps $\mu : C \to X$ such that $\mu_*[C] = \beta$.

The construction of $\overline{M}_{0,n}(X, \beta)$ is a little harder: you first embed X into a big projective space $\mathbb{P}^r$ and then relate to the previous construction. Once it has been constructed, the moduli space enjoys the same properties as in the case $\overline{M}_{0,n}(\mathbb{P}^r, d)$. The dimension is

$$\dim X + n - 3 + \int_\beta c_1(T_X). \tag{2.10.1.1}$$

It is not known in general whether $\overline{M}_{0,n}(X, \beta)$ is irreducible. Irreducibility has been established only for generalized flag manifolds, i.e., spaces of the form G/P (cf. [79], [48]).

2.10.2 More general varieties. For general smooth projective varieties X, it is still true that there exists a coarse moduli space $\overline{M}_{0,n}(X, \beta)$, and it is in fact projective. But in general it is not irreducible or connected, and it will typically have components of excessive dimension, that is, greater than the expected dimension given in 2.10.1.1.

Often the "boundary" is of higher dimension than the locus of irreducible maps, so strictly speaking $\overline{M}_{0,n}(X, \beta)$ cannot really be considered a compactification of $M_{0,n}(X, \beta)$. The simplest example is $X = \widetilde{\mathbb{P}}^2 \xrightarrow{\varepsilon} \mathbb{P}^2$, the blowup of $\mathbb{P}^2$ at a point

q. Let h denote the class of the pullback of a line from $\mathbb{P}^2$, and let e denote the class of the exceptional divisor E. Now consider the class $\beta = 4h \in A_1(X)$. The expected dimension of $\overline{M}_{0,0}(X, \beta)$ is 11. In fact, in the locus of irreducible maps, all maps are disjoint from E, so the natural morphism $M_{0,0}(X, \beta) \to M_{0,0}(\mathbb{P}^2, 4)$ (composition with ε) is an isomorphism. Now we claim that $\overline{M}_{0,0}(X, \beta)$ contains a "boundary divisor" D of dimension 12: it consists of maps such that one twig maps to a curve that meets E three times, and one twig is a triple cover of E. For the dimension count, admit the fact that there is an 8-dimensional family of rational plane quartics with a triple point at q. The strict transforms of these curves show that there are also 8 dimensions in $M_{0,0}(X, 4h - 3e)$, and all curves herein meet E in three points (counted with multiplicity). On the other hand, since $E \simeq \mathbb{P}^1$, we see that $M_{0,0}(E, 3e)$ (the triple covers of E) has dimension 4. Now D is obtained by gluing these triple covers to the curves of degree $4h - 3e$.

It can also happen that the locus of irreducible curves is empty while the boundary is not! Let Y be the blowup of X at a point on E. Then the union of the two exceptional curves is a reducible genus-0 curve, and it is the only curve in its homology class β. So $M_{0,n}(Y, \beta)$ is empty! But the "compactification" $\overline{M}_{0,n}(Y, \beta)$ is nonempty, and all the maps in it have reducible source. (Although this last example may look artificial, it nevertheless plays a nontrivial role in Graber [37] § 3.3.)

Also, irreducible maps can form components of too-high dimension, so even $M_{0,n}(X, \beta)$ (no bar) can be ill-behaved: this happens in connection with multiple-cover maps. A famous example is the general quintic threefold $Q \subset \mathbb{P}^4$. Since the c_1 of the tangent bundle of Q is trivial, the expected dimension of $M_{0,0}(Q, d)$ is zero, independent of the degree d (d means d times the class of a line). It is known that there are 2875 lines on Q, and 609250 smooth conics. Now for each of these smooth conics there is an irreducible zero-dimensional component (an isolated point!) of $M_{0,0}(Q, 2)$, and in addition to that for each line there is an irreducible component of $M_{0,0}(Q, 2)$ consisting of double covers of the line. Each of these double-cover components is isomorphic to $M_{0,0}(\mathbb{P}^1, 2)$, which is of dimension 2. So altogether $M_{0,0}(Q, 2)$ has 612125 components, and 2875 of them are of too-high dimension!

The quintic threefold is perhaps the single most important variety to count rational curves in, due to the central role it plays in *mirror symmetry*; see the book of Cox and Katz [13] for an introduction to this hot topic, and for details on the quintic threefold in particular.

2.10.3 Positive genus. The complications in the case of positive genus include all those described in the end of the first chapter, since curves of equal genus are not necessarily isomorphic; reducible curves are not necessarily trees, etc. It is true that there exists a projective coarse moduli space $\overline{M}_{g,n}(X, \beta)$, but there occur phenomena similar to those in the case of a nonconvex target space: the moduli spaces are in general reducible and have components of too-high dimension.

A simple example is $\overline{M}_{1,0}(\mathbb{P}^2, 3)$. Its locus of irreducible maps is birational to the $\mathbb{P}^9$ of plane cubics, and thus of dimension 9. But in addition to this good component (the closure of this locus), there is a component of excessive dimension, namely the "boundary" component consisting of maps having a rational twig of degree 3 and an elliptic twig contracting to a point. The dimension of this component is

$$\dim \overline{M}_{0,1}(\mathbb{P}^2, 3) + \dim \overline{M}_{1,1} = 9 + 1 = 10.$$

(The marks on these two spaces are the gluing marks.)

2.10.4 Deformation theory. A compulsory next step in the study of these moduli spaces is to do some rudimentary deformation theory. Let us briefly touch upon the most basic notions; a good reference is Harris–Morrison [43], Section 3B. For simplicity we consider the case of no marks. Let $\mu : C \to X$ be a general point of $\overline{M} = \overline{M}_{g,0}(X, \beta)$, so we assume that C is smooth and that μ is an immersion. Then there is a well-behaved normal bundle of μ, denoted by N_μ, defined by the short exact sequence

$$0 \to T_C \to \mu^* T_X \to N_\mu \to 0. \qquad (2.10.4.1)$$

There is an isomorphism (the Kodaira–Spencer map) between the tangent space of $\overline{M}$ at μ and $H^0(C, N_\mu)$, the space of first-order infinitesimal deformations of μ. Assuming furthermore that the first-order deformations are unobstructed, (e.g., $H^1(C, N_\mu) = 0$) we can compute the dimension of $\overline{M}$ at μ by Riemann–Roch (cf. Fulton [28], Ex. 15.2.1). The result is

$$\dim \overline{M}_{g,0}(X, \beta) = h^0(C, N_\mu) = (\dim X - 3)(1 - g) + \int_\beta c_1(T_X).$$

This is what is called the *expected dimension* of $\overline{M}_{g,0}(X, \beta)$, and if there are n marks, clearly the dimension is n higher.

2.10.5 Stacks. We refer the reader either to the notes of Edidin [20], a nice introduction to moduli of curves that adopts the language of stacks, or to the short *Stacks for Everybody* by Fantechi [26]. (References to the heavier literature can be found in these two.)

Here we just give a bare outline of a few of the central ideas.

Deligne–Mumford stacks are geometric objects that look like schemes locally in the étale topology. From the viewpoint of moduli theory, stacks generalize schemes by incorporating information about all the automorphisms of the objects parametrized. It is just the concept needed to make the theory of moduli much smoother, literally: for instance, the stack version of $\overline{M}_{0,n}(\mathbb{P}^r, d)$ is smooth and possesses a universal family, naturally identified with (the stack version of) $\overline{M}_{0,n+1}(\mathbb{P}^r, d)$.

Many problems and peculiar phenomena due to the presence of automorphisms, encountered in Section 2.6 and in the construction of families in $\overline{M}_{0,0}(\mathbb{P}^2, 2)$ in Section 2.9, are more naturally dealt with in the language of stacks. The trouble can be traced back to the very definition of moduli functor, which involves dividing out by isomorphisms. In some situations (in the presence of nontrivial automorphisms) this is too crude an approach.

The idea of stacks is simply not to divide out by those isomorphisms! Instead of just looking at the *set* of isomorphism classes of families, look at the *groupoid* of all the families themselves *and* the isomorphisms between them.

GROUPOIDS. A *groupoid* is a category all of whose arrows are isomorphisms. This generalizes the notion of equivalence relation: an equivalence relation is essentially a groupoid in which there is at most one arrow connecting any pair of objects (then two objects are related if and only if they are connected by an arrow). Given a groupoid G, one can consider $\pi_0(G)$, the set of isomorphism classes of its objects (this generalizes the quotient of an equivalence relation). Groupoids also generalize sets: a set is just a groupoid in which the only arrows are the identity arrows. Let ***Grpd*** denote the category of all groupoids (and functors between them).

Now instead of considering a moduli functor $\boldsymbol{Sch}^{\mathrm{op}} \to \boldsymbol{Set}$ sending a scheme B to the set of isomorphism classes of families $\mathfrak{X}/B$, consider a "functor" $\boldsymbol{Sch}^{\mathrm{op}} \to \boldsymbol{Grpd}$, sending B to the groupoid whose objects are families $\mathfrak{X}/B$ and whose arrows are isomorphisms between such families, and sending a scheme morphism $\varphi : B' \to B$ to the pullback functor φ^*. We see that the old set-valued functor is just the π_0 of this groupoid-valued one.

PSEUDOFUNCTORS. The groupoid-valued functor is in quotes because strictly speaking it is not a functor: since pullbacks are only defined up to isomorphism, we get only a functor up to isomorphism. Such a thing is called a *pseudofunctor* if the comparison isomorphisms are sufficiently coherent, e.g., come from a universal property as in the case of pullbacks.

THE ÉTALE TOPOLOGY. A morphism of schemes is *étale* if it is flat and unramified. In particular, each fiber is a smooth 0-dimensional scheme. The *étale topology* is an example of a *Grothendieck topology*: it is specified not merely in terms of open subsets and their intersections but more generally in terms of étale morphisms and their fiber products.

STACKS. A pseudofunctor $F : \boldsymbol{Sch}^{\mathrm{op}} \to \boldsymbol{Grpd}$ is called a *stack* if it satisfies two gluing conditions that together form the groupoid analogue of the sheaf condition for set-valued functors. The conditions roughly say that we can patch local data to get global objects if the local data "agree" on "overlaps" (in the sense of the étale topology). There is one condition for objects (families) and one for arrows (isomorphisms between families). For arrows, the condition is just a sheaf condition, meaning that isomorphisms glue if they match locally. For objects, the condition is *effectivity of descent*: it is roughly the sheaf condition again, but since we are in a situation in which the objects are not the highest level of structure—there are also arrows (which happen to be isomorphisms)—the sheaf condition is only up to isomorphism, in a certain precise sense. This means that objects can be glued if just locally they match up to coherent isomorphisms (i.e., satisfying a cocycle condition). We will not go into the precise definitions; the reader can find the details in the literature, starting with [26] or [20].

REPRESENTABLE MORPHISMS. The functor of points of a scheme Y can also be interpreted as a pseudofunctor $\boldsymbol{Sch}^{\mathrm{op}} \to \boldsymbol{Grpd}$, and it is in fact a stack which we also denote by Y. Now one can say that a stack F is *representable* if it is equivalent to a scheme in this sense, and a morphism of stacks $F \to G$ is called *representable* if for every scheme Y and morphism $Y \to G$, the fiber product[2] $Y \times_G F$ is representable.

In this way, many notions of scheme theory make sense also for stacks. In particular, all properties of morphisms of schemes that are stable under base change and of a local nature on the codomain have a meaning for representable morphisms of stacks: such a property is said to hold for a representable morphism of stacks if it holds for every base change to a scheme.

DELIGNE–MUMFORD STACKS. A stack F is *Deligne–Mumford* if every morphism from a scheme $Y \to F$ is representable, quasicompact, and separated, and if there exists a scheme U and a surjective étale morphism of stacks $U \to F$.

The last condition says that F can be covered by a scheme in the étale topology; this scheme U should be thought of as an atlas of F. In practice, this is the key point: really a lot of geometry makes sense for Deligne–Mumford stacks, by working on the atlas; in particular, smooth Deligne–Mumford stacks have a good intersection theory.

Not every moduli problem admits a Deligne–Mumford moduli stack, but if there is a coarse moduli space, typically there will be a smooth Deligne–Mumford stack. The relation between the two notions is this: a coarse moduli space for a Deligne–Mumford stack F is a pair (M, v), where M is a scheme and $\mathsf{v} : F \to M$ is a proper morphism of stacks such that: (i) (M, v) is initial among such pairs, (ii) there is a bijection of sets $\pi_0(\mathsf{v}_\bullet) : \pi_0 F(\bullet) \to \pi_0 M(\bullet) = M(\bullet)$. Compare with Definition 0.2.9.

Note that there are two "scheme approximations" involved: the covering scheme U and the "primitive scheme quotient" M. It is fruitful to think also of F as a sort of quotient of U, called the *stack quotient*, by construction a very good quotient, since $U \to F$ is étale.

---

[2]The correct fiber product here is not the naive one like "pairs-of-objects-with-equal-image," but rather a 2-categorical notion like "pairs-of-objects-with-an-isomorphism-between-the-images."

STACKS OF STABLE MAPS. The stack of stable maps $F = \overline{M}_{0,n}(\mathbb{P}^r, d)$ is a smooth and proper Deligne–Mumford stack. We mentioned in 2.4.8 that the coarse moduli scheme of stable maps $M = \overline{M}_{0,n}(\mathbb{P}^2, d)$ is locally a quotient Y/G of a smooth scheme by a finite group of order $d!\,d!\,d!$, and the global M is obtained by gluing together those Y/G. The stack of stable maps F can be seen as a better version of this quotient. In this case the atlas U is obtained by gluing together the smooth schemes Y, and then $U \to F$ is an étale map of degree $d!\,d!\,d!$.

In Example 2.9.9 we had a 1-parameter family of maps where we had to make a base change $b \mapsto b^2$ in order to get a stable map in the $(b=0)$-fiber. You could say that we were actually working on the étale 2:1 cover of the stack version of the family. The original base B (the affine line, target of $b \mapsto b^2$) was a baby model of the coarse moduli space, a bad quotient, and having the defect of not carrying a family. The stack version S of the base, which was not made explicit, sits in between the two affine lines, and it has a sort of magic point at 0 (often called a *stacky point*), making the map "$b \mapsto b^2$" étale:

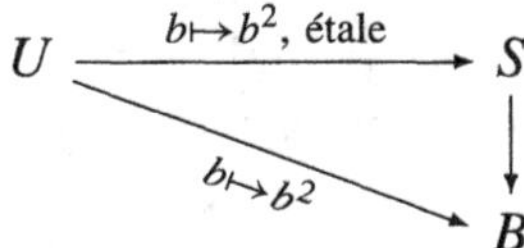

2.10.6 Readings. The reader may (should?) study all the details of the construction of the moduli space $\overline{M}_{0,n}(\mathbb{P}^r, d)$ (or more generally, $\overline{M}_{0,n}(X, \beta)$, with X convex) in the first six sections of FP-NOTES.

However, we wish to make the point that a lot of good geometry can be done *assuming* the existence and the basic properties of $\overline{M}_{0,n}(\mathbb{P}^r, d)$.

We recommend the excellent and accessible paper of Pandharipande [69]. First, a description of natural generators of the Picard group $\mathrm{Pic}(\overline{M}) \otimes \mathbb{Q}$ is given. In general, $\mathrm{Pic}(\overline{M}) \otimes \mathbb{Q}$ is generated by all the boundary divisors together with $\nu_i^*(H)$ (inverse image of the hyperplane class H). (For $n \geq 3$, these inverse image divisors can be substituted by the incidence divisor $\mathrm{inc}(H^2) = \varepsilon_* \nu_{n+1}^*(H^2)$ (cf. 2.6.4).) Next, 1-parameter families are exploited systematically to express the class of various divisors of interesting geometric meaning in terms of the generators. This leads to an algorithm for computing characteristic numbers of rational curves in $\mathbb{P}^r$ (cf. 3.6.2).

Let us finally mention the papers of Vakil [83] and [84], which treat the spaces $\overline{M}_{g,0}(\mathbb{P}^2, d)$ for $g = 1, 2, 3$, using techniques similar to those of [69].

Exercises

1. Consider the punctured family of parametrizations of twisted cubics in $\mathbb{P}^3$ given by

$$\begin{array}{ccc} B \times \mathbb{P}^1 & \dashrightarrow & \mathbb{P}^3 \\ \left(b, \begin{bmatrix} s \\ t \end{bmatrix}\right) & \longmapsto & \begin{bmatrix} bs^3 \\ s^2t \\ st^2 \\ t^3 \end{bmatrix} \end{array}.$$

Blow up the point of indeterminacy to resolve the map, plombing the family. Show that the limit curve is the union of the plane conic with equations $0 = X_0 = X_2^2 - X_1X_3$ and the line with equations $0 = X_2 = X_3$.

Background on dual curves and parametrizations of duals of rational curves

2. Let C be a smooth plane curve given by an equation $F(X_0, X_1, X_2) = 0$. Since C is smooth, the three partial derivatives $F_i = \frac{\partial F}{\partial X_i}$ do not vanish simultaneously at any point. For each point $P \in C$, show that the line given by equation

$$F_0(P)X_0 + F_1(P)X_1 + F_2(P)X_2 = 0$$

is the tangent line to C at P. This defines the Gauss map of C,

$$\begin{array}{ccc} C & \longrightarrow & \check{\mathbb{P}}^2 \\ P & \longmapsto & [F_0(P) : F_1(P) : F_2(P)]. \end{array}$$

3. Find the image of the Gauss map for the plane conic C given by the equation $XY - Z^2$.

4. Consider a degree-d immersion

$$\begin{array}{ccc} \varphi : \mathbb{P}^1 & \longrightarrow & \mathbb{P}^2 \\ \begin{bmatrix} s \\ t \end{bmatrix} & \longmapsto & \begin{bmatrix} \varphi_0(s,t) \\ \varphi_1(s,t) \\ \varphi_2(s,t) \end{bmatrix} \end{array}.$$

(i) For any point $p \in \mathbb{P}^1$, show that the corresponding tangent line of the image curve can be parametrized by the linear map $\mathbb{P}^1 \to \mathbb{P}^2$ given by the Jacobian matrix of φ at p. To get the equation of this tangent, and thus the corresponding point in the dual plane $\check{\mathbb{P}}^2$, pick an "orthogonal vector," say the cross product of the columns of the matrix.

(ii) Show that the cross product of the columns of the Jacobian matrix provides a parametrization of the dual curve. Conclude from this description that the dual curve has degree $2d-2$. (This also follows from the Plücker formula: if C is a smooth plane curve of degree d then the dual has degree $d(d-1)$, and if C has δ double points, then the degree of the dual is 2δ less. In the case of an immersion of degree d there are $(d-1)(d-2)/2$ double points, so the degree of the dual curve is $d(d-1)-(d-1)(d-2)=2d-2$.)

5. Consider the following family of parametrized rational cubics:

$$\begin{array}{rcl} \mu_b : \mathbb{P}^1 & \longrightarrow & \mathbb{P}^2 \\ \begin{bmatrix} s \\ t \end{bmatrix} & \longmapsto & \begin{bmatrix} st^2 - bs^3 \\ t^3 - bs^2 t \\ s^3 \end{bmatrix}. \end{array}$$

The general fiber is a nodal cubic, for example, for $b=1$, the equation of the image is $X^3+X^2Z-Y^2Z$, which is one of the most popular nodal cubics, whereas for $b=0$ we get X^3-Y^2Z, which we recognize as our favorite cuspidal cubic: the cusp is at $[\begin{smallmatrix}1\\0\end{smallmatrix}] \mapsto \begin{bmatrix}0\\0\\1\end{bmatrix}$, and the inflection point is at $[\begin{smallmatrix}0\\1\end{smallmatrix}] \mapsto \begin{bmatrix}0\\1\\0\end{bmatrix}$, the flex line being $Z=0$.

(i) Using Exercise 4 on the facing page, show that the dual family of parametrizations is given by

$$[-3s^2(3t^2-bs^2) \;:\; 6s^3t \;:\; (t^2-3bs^2)(3t^2-bs^2)+4bs^2t^2].$$

(For $b \neq 0$, this is the rational quartic with three cusps given by the equation

$$27Y^2Z^2+4X^3Z-4bX^4-36bXY^2Z+8b^2X^2Y^2-4b^3Y^4,$$

as you can verify by plugging the parametrization into the equation. Unless you can do this by machine, don't bother.)

(ii) Show that the $(b{=}0)$-member of the family,

$$[-9s^2t^2 \;:\; 6s^3t \;:\; 3t^4]$$

is a rational map parametrizing a cuspidal cubic (equation $27Y^2Z+4X^3$), with cusp at $[\begin{smallmatrix}0\\1\end{smallmatrix}] \mapsto \begin{bmatrix}0\\0\\1\end{bmatrix}$, and inflection point at $[\begin{smallmatrix}1\\0\end{smallmatrix}] \mapsto \begin{bmatrix}0\\1\\0\end{bmatrix}$. Compare with the cuspidal cubic on the other side of the duality.

(iii) The parametrization in (ii) is not defined for $t=0$. Blow up at this point to resolve the map. Show that on the strict transform of the $(b{=}0)$-fiber you find the everywhere-defined parametrization of the above cuspidal cubic, and on the exceptional divisor you get the parametrization of the line Z, which is exactly the pencil of lines through the origin of the original plane.

Summing up, the dual family has as general member a map from $\mathbb{P}^1$ to a quartic with three cusps. The special member is a map from a two-component curve to the union of a cuspidal cubic and its inflectional tangent.

6. The family of parametrized cubics

$$\begin{aligned} \mu : \mathbb{P}^1 & \longrightarrow \mathbb{P}^2 \\ \begin{bmatrix} s \\ t \end{bmatrix} & \longmapsto \begin{bmatrix} 3b^2st^2 \\ t^3 - 3s^2t \\ s^3 \end{bmatrix} \end{aligned}$$

satisfies the family of equations

$$X^3 - 18b^2X^2Z + 81b^4XZ^2 - 27b^6Y^2Z$$

(as you might check just by substituting the parametrization into the equation). For $b \neq 0$, the cubics are nodal (with a double point on the line $Y = 0$).

(i) Show that for $b = 0$, we get a 3-fold covering of the line $X = 0$, with two simple branch points and one double branch point.

(ii) The dual curve of the general member of the family is a quartic with three cusps. Argue why the limit of those quartics should be a 3-component curve in $\mathbb{P}^2$: two simple lines and one double line.

(iii) Show that the guess in the previous item is correct, by constructing the dual family of parametrizations as in the previous two exercises. The dual family has three points of indeterminacy in the central fiber $b = 0$, all three in the $(t = 1)$-chart. These points are $s = 0$, $s = \pm 1$. (These three points correspond exactly to the branch points of the special fiber on the other side of the duality.) Blow up each of the three points and find in each case that the exceptional divisor maps to a line in $\mathbb{P}^2$: this line is the pencil of lines through the branch points on the other side of the duality. In the $(s = 0)$-point we get a double cover of a line (reflecting the fact that the corresponding branch point is double).

7. Modify the initial parameters of the previous exercise to obtain instead a family of nodal cubics degenerating into a 3-fold cover with four simple branch points.

Complete conics—some more background for Section 2.9. We will employ matrix notation for conics, so let us make explicit that our projective plane is $\mathbb{P}^2 = \mathbb{P}E$

where E is the space of column vectors $\mathbf{x} = \begin{bmatrix} x_0 \\ x_1 \\ x_2 \end{bmatrix}$. We denote by $[\mathbf{x}]$ the corresponding point in $\mathbb{P}E$.

The variety of all plane conics can be described as $\mathbb{P}^5 = \mathbb{P}(\mathrm{Sym}^2(E))$, the projectivization of the space of symmetric 3×3 matrices: a matrix A defines a conic (also denoted by A) by the equation $\mathbf{x}^T A\mathbf{x} = 0$.

The variety $W \subset \mathbb{P}^5$ of singular conics (line pairs) is the locus of matrices of rank ≤ 2, and the variety of double lines $V \subset \mathbb{P}^5$ is the locus of matrices of rank 1. It is the image of the Veronese embedding $\check{\mathbb{P}}^2 \to \mathbb{P}^5$ sending a linear form to its square.

8. (i) Show that the tangent space of A at a smooth point $[\mathbf{x}]$ is the set of $[\mathbf{y}]$ such that $\mathbf{x}^T A\mathbf{y} = 0$, and conclude that the Gauss map is described as

$$\begin{array}{ccc} A & \longrightarrow & \check{\mathbb{P}}^2 \\ [\mathbf{x}] & \longmapsto & [\mathbf{x}^T A]. \end{array}$$

(ii) Show that if A is smooth (i.e., of rank 3), the image conic is smooth again and is given by the matrix A^{-1}, or equivalently, since we only consider matrices up to scalar multiplication, the cofactor matrix A' (recall that $A^{-1} = A'/\det A$).

This defines the rational map

$$\begin{array}{cccc} \varphi : & \mathbb{P}^5 & \dashrightarrow & \check{\mathbb{P}}^5 \\ & A & \longmapsto & A'. \end{array}$$

This map is also well defined on W since the cofactor matrix of a rank-2 matrix is well defined (it is a matrix of rank 1). In contrast, the cofactor matrix of a rank-1 matrix is the zero matrix, which we do not allow here, so φ is not defined on V.

(iii) Given a rank-2 matrix A corresponding to a line pair $L + M \subset \mathbb{P}^2$. Show that cofactor matrix A' (rank 1) corresponds to the double line Q in $\check{\mathbb{P}}^2$ that is the pencil of lines though the intersection point $L \cap M$ in $\mathbb{P}^2$.

9. The variety of complete conics $\mathbb{B} \subset \mathbb{P}^5 \times \check{\mathbb{P}}^5$ is by definition the closure of the graph of φ.

(i) Show that $\mathbb{B}$ can be described as the variety of pairs of symmetric matrices (A, B) such that $3AB = \mathrm{trace}(AB)I$.

(ii) Show that $\mathbb{B}$ is the blowup of $\mathbb{P}^5$ along V. (*Hint:* the ideal of V is generated by the cofactors.) Show that the exceptional divisor is naturally isomorphic to $\mathbb{P}^2 \times \mathbb{P}^2$ blown up along the diagonal.

(iii) Verify the following case-by-case description of $\mathbb{B}$, which does not make reference to the dual conics: a points in $\mathbb{B}$ is either

— a smooth conic

— a line pair

— a double line with a distinguished degree-2 divisor (i.e., two points, possibly coincident).

These two points are called the *foci*.

10. Show that W is the secant variety of V. (*Hint:* a linear combination of two rank-1 matrices has rank at most 2.)

11. *Limit interpretation of foci.* Let $C_{[a:b]}$ be a pencil of conics whose general member is a smooth conic.

    (i) Show that $C_{[a:b]}$ contains at most one double line. (*Hint:* use the previous exercise.)

    (ii) Show that if $C_{[a:b]}$ contains a double line L then the pencil has two base points (i.e., points common to all members of the pencil). (The two base points might coincide: this happens if all the members of the pencil are tangent to L.)

    (iii) Show that the dual pencil contains a line pair, namely the two lines dual to the base points. In particular, in terms of complete conics, the foci of L are precisely the base points of the family. (In the case where the two base points coincide, the limit line pair in the dual family is actually a double line.)

12. Use matrix machinery to describe the family of 2.9.9 and verify the claim made there that the dual family is a double line. Identify the foci (necessarily coincident) for both the family and its dual. (Check also that the base change does not alter any of the arguments involved.)

13. (i) Identify the foci of the $a = 0$ limit of the pencil $aXY - bZ^2$. (This is the other end of the pencil considered in Example 2.2.1.)

    (ii) The same pencil can be described parametrically:

$$\left(\begin{bmatrix} a \\ b \end{bmatrix}, \begin{bmatrix} s \\ t \end{bmatrix}\right) \mapsto \begin{bmatrix} bs^2 \\ at^2 \\ ast \end{bmatrix}$$

    but there is a point of indeterminacy at $(a, s) = (0, 0)$. Show that although blowing up a couple of times does resolve the map, the resulting $a = 0$ member of the family is not a stable map. (It is a degree-1 map from a nonreduced fiber.)

14. (Base change of Exercise 13 on the preceding page.) (i) In the 1-parameter family of conics given by $a^2XY - b^2Z^2$, identify the foci of the $a = 0$ member (the double line Z^2.)

    (ii) The same family can be described in terms of parametrizations as

$$\left(\begin{bmatrix} a \\ b \end{bmatrix}, \begin{bmatrix} s \\ t \end{bmatrix}\right) \mapsto \begin{bmatrix} bs^2 \\ bt^2 \\ ast \end{bmatrix}.$$

    (There is a point of indeterminacy for $b = 0$, but we are only concerned with $a = 0$, and here the map is well defined.) For $a = 0$ we get a double cover of the line $Z = 0$. Show that the branch points are precisely the foci; cf. item (i).

Background material on $G = G(2, 4)$, the Grassmannian of lines in $\mathbb{P}^3$

15. *Reminders on the tangent bundle of G.* Let $0 \to S \to \mathbb{C}^4 \to Q \to 0$ be the tautological exact sequence of vector bundles over G. The fiber of S over $\ell \in G$ is the corresponding two-dimensional vector subspace of $\mathbb{C}^4$. The tangent bundle TG is isomorphic to $\mathrm{Hom}(S, Q) \simeq S^\vee \otimes Q$ as shown in the following steps (cf. [2], Prop. 2.7).

    (i) Let $X \to Y$ be smooth. Then the relative tangent bundle TX/Y is naturally isomorphic to the normal bundle of the diagonal embedding $X \hookrightarrow X \times_Y X$.

    (ii) Let $E \to X$ be a vector bundle of rank r and let s be a regular section, i.e., in a suitable local trivialization, it is expressed by a vector $(f_1, \dots, f_r)$ of functions that form a regular sequence: each f_i is not a zero divisor modulo the ideal, $\langle f_1, \dots, f_{i-1}\rangle$, generated by the previous ones. Let $Z \subset X$ be the scheme of zeros of S (so that the local ideal of Z is $\langle f_1, \dots, f_n\rangle$). Then the normal bundle of Z in X is naturally isomorphic to the restriction E_Z.

    (iii) Let $p_i : G \times G \to G$ be the projections. Take the pullback of the tautological sequence and let s be the section of $\mathrm{Hom}(p_1^*S, p_2^*Q)$ defined by composing the homomorphism $p_1^*S \to \mathbb{C}^4 \to p_2^*Q$. Show that for any map $\ell : X \to G \times G$ the following are equivalent: (1) $p_1 \circ \ell = p_2 \circ \ell$; (2) ℓ factors through the diagonal; (3) $\ell^*s = 0$. Hence the diagonal $\Delta \subset G \times G$ is the scheme of zeros of s and $TG \simeq \mathrm{Hom}(p_1^*S, p_2^*Q)_\Delta \simeq \mathrm{Hom}(S, Q)$, since $p_1 = p_2$ over $G \simeq \Delta$.

    (iv) Compute the first Chern class of TG in terms of $c_1 Q$.

16. (i) Show that the tangent bundle of G is generated by global sections. (*Hint:* use the surjection $(\mathbb{C}^4)^\vee \otimes \mathbb{C}^4 \to S^\vee \otimes Q \simeq TG$.)

    (ii) Show that G is convex (cf. 2.10.1: for every map $\mu : \mathbb{P}^1 \to G$ we have $H^1(\mathbb{P}^1, \mu^*T_G) = 0$). *Hint:* Any vector bundle over $\mathbb{P}^1$ is isomorphic to

$\bigoplus \mathcal{O}(d_i)$ for suitable integers d_i; this is globally generated if and only if all $d_i \geq 0$.)

17. By the previous exercise, and by the general result quoted in 2.10.1, there is a coarse moduli space $\overline{M}_{0,n}(G, d)$ of stable genus-0 maps to G of degree d. Here d means d times the class of a 1-dimensional Schubert variety (of lines contained in a plane and passing through a point in that plane). Use formula 2.10.1.1 to compute the dimension of $\overline{M}_{0,n}(G(2, 4), d)$.

Two friendly exercises as dessert

18. Generalizing Exercise 9 on page 45, show that for $n > 0$, the number of boundary divisors of $\overline{M}_{0,n}(\mathbb{P}^r, d)$ is
$$2^{n-1}(d+1) - n - 1$$
and that the number of boundary divisors of $\overline{M}_{0,0}(\mathbb{P}^r, d)$ is $[d/2]$, the integral part of $d/2$.

19. Show that the same formulas hold for every smooth convex variety X whose A_1 is of rank 1 (e.g., the Grassmannian of lines in $\mathbb{P}^3$), if d is interpreted as meaning d times the positive generator for $A_1(X)$.

Chapter 3

Enumerative Geometry via Stable Maps

3.1 Classical enumerative geometry

3.1.1 The principle of conservation of number. Classical approaches to enumerative questions frequently made use of the "principle of conservation of number." Roughly speaking, it was tacitly assumed that the number of solutions to a counting problem remains constant when the "generic" conditions imposed are moved to special position.

A typical example is the determination of the number of lines in $\mathbb{P}^3$ incident to four lines $\ell_1, \dots, \ell_4$ in general position. Specializing the lines in such a way that ℓ_1, ℓ_2 intersect at a point p, and ℓ_3, ℓ_4 intersect at another point q, we see that there are exactly two solutions: one is the line $\overline{pq}$, and the other is the line along which the two planes $\langle \ell_1, \ell_2 \rangle$ and $\langle \ell_3, \ell_4 \rangle$ intersect.

Of course it is necessary to justify not only the conservation of number but the noninterference of multiplicities as well. The need of a critical revision of classical enumerative geometry, establishing the limits of validity of the methods and results of Schubert and his school, was formulated by Hilbert as the fifteenth problem of his famous list presented at the meeting of the International Mathematics Union at the turn of the century. See the survey of Kleiman [50].

3.1.2 Enumerative geometry via intersection theory. The most successful post-classical approaches consist in applying well-established methods of intersection theory to parameter spaces set up for each specific counting problem. The idea is simple: the family of objects are put in one-to-one correspondence with the points of an algebraic variety M (the parameter space or moduli space), and each condition then cuts out a subvariety in M. Thus the object satisfying all the conditions corresponds to the points in the intersection of these subvarieties. In this way the enumeration problem is turned into a question of counting points in an intersection of algebraic varieties, i.e., a problem in intersection theory.

In the above example, one can work in the Grassmannian $\mathrm{Gr}(1, \mathbb{P}^3)$ of lines in $\mathbb{P}^3$. The Plücker embedding realizes $\mathrm{Gr}(1, \mathbb{P}^3)$ as a quadric hypersurface in $\mathbb{P}^5$. The condition of incidence to a line ℓ_i is given as the intersection of the quadric $\mathrm{Gr}(1, \mathbb{P}^3) \subset \mathbb{P}^5$ with its embedded tangent hyperplane at the point $[\ell_i] \in \mathrm{Gr}(1, \mathbb{P}^3)$. In general, the intersection of the four hyperplanes defines a line in $\mathbb{P}^5$, which by Bézout intersects $\mathrm{Gr}(1, \mathbb{P}^3)$ in two points (possibly coincident).

In order for this approach to work in general, a compact parameter space is required together with some knowledge of its intersection ring. Often the original parameter space M is not compact; one has to find a compactification $\overline{M}$, and in the end of the count it must be checked that the found solutions are in fact in the dense open set M corresponding to the original objects.

Second, the knowledge of the intersection ring of $\overline{M}$ comes from the geometry of $\overline{M}$, which in turn is described in terms of the geometry of the objects it parametrizes. For this to work it is crucial that the points in $\overline{M} \smallsetminus M$ (the boundary) can be given geometric interpretation as well, typically as degenerations of the original objects (points in M). In other words, the compactification too must be a parameter space of something; this is called a modular compactification.

3.1.3 Example: plane conics. *How many smooth conics pass through* 5 *points in general position in the plane?* The space of smooth conics is an open set U in the $\mathbb{P}^5$ of all homogeneous polynomials of degree 2: to each conic is associated the coefficients of its equation (up to the multiplication of a nonzero scalar). We can simply take $\mathbb{P}^5$ as our compactification. The condition of passing through a given point corresponds to a hyperplane in $\mathbb{P}^5$. Since the points are assumed to be in general position, the intersection of five such hyperplanes constitutes a unique solution; it remains to check that this point is in U. This follows from a geometric argument: the points in $\mathbb{P}^5 \smallsetminus U$ correspond to line pairs and double lines, and no such configuration can pass through 5 points, unless three of the points are collinear (and thus not in general position).

The same reasoning holds for the count of plane cubics passing through 9 points, or more generally, plane curves of degree d passing through $d(d+3)/2$ points; in each case the answer is 1.

3.1.4 Example: rational cubics. The situation changes when we ask for the number of *rational* plane cubics. A rational plane cubic is necessarily singular. The singular cubics form the discriminant hypersurface D in the $\mathbb{P}^9$ of all the cubics. The pertinent question is how many rational cubics pass through 8 points: this corresponds to intersecting D with 8 hyperplanes. The number of points in this intersection is 12, the degree of the hypersurface D. The discriminant is a special case of the notion of dual variety. Its degree can be computed as in Fulton [28].

A topological argument for the present count is roughly this. We want to count the points of intersection of D with a general line in $\mathbb{P}^9$. This line constitutes a pencil of plane cubics $\{t_1 F_1 + t_2 F_2\}_{[t_1:t_2]\in\mathbb{P}^1}$, where F_1, F_2 are two general cubics. Blowing up the 9 points of intersection of these two cubics, we obtain a surface S and a morphism $t : S \to \mathbb{P}^1$ that extends the rational map $\mathbb{P}^2 \dashrightarrow \mathbb{P}^1$ defined by the pencil:

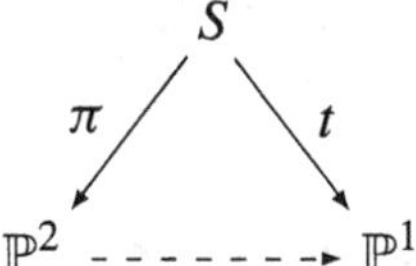

The fiber of t over $[t_1 : t_2] \in \mathbb{P}^1$ is isomorphic to the curve given by $t_1 F_1 + t_2 F_2$ (since none of the 9 base points are singular points); for most values of $[t_1 : t_2]$ this is a smooth curve of genus 1, but for a finite set $\Sigma \subset \mathbb{P}^1$ the fibers are singular curves, isomorphic to a nodal plane cubic. We want to compute the cardinality n of Σ. So topologically, the restriction of t to $U = \mathbb{P}^1 \smallsetminus \Sigma$ is a torus fibration (let $\mathbb{T}$ denote a torus), and the fibers over Σ are "pinched" tori $\mathbb{S}$. Using the properties of the topological Euler characteristic we can compute, on the one hand,

$$\begin{aligned}\chi(S) &= \chi(\mathbb{P}^2 \smallsetminus \{p_1, \dots, p_9\}) + 9\,\chi(\mathbb{P}^1) \\ &= 3 - 9 + 18 = 12,\end{aligned}$$

and on the other hand,

$$\begin{aligned}\chi(S) &= \chi(t^{-1}U) + \chi(t^{-1}\Sigma) \\ &= \chi(U)\cdot\chi(\mathbb{T}) + n\cdot\chi(\mathbb{S}).\end{aligned}$$

The first part of this last expression is zero since $\chi(\mathbb{T}) = 0$. On the other hand, since $\chi(\mathbb{S}) = 1$, we conclude that $n = 12$. See the survey of Caporaso [11] for other ways of computing specifically this number.

3.1.5 Higher degree. As the degree d increases, the situation becomes more complicated. Recall the genus formula for a nodal plane curve,

$$g = \frac{(d-1)(d-2)}{2} - \delta,$$

where δ is the number of nodes. Thus, to get rational curves we must impose $(d-1)(d-2)/2$ nodes. It is a fact that each node is a condition of codimension 1; that is, in the space $V \subset \mathbb{P}^{d(d+3)/2}$ of all irreducible curves of degree d, the rational ones constitute a subvariety V_0^d of dimension

$$\dim V_0^d = d(d+3)/2 \;-\; (d-1)(d-2)/2 = 3d - 1.$$

To get a finite number of curves we must impose $3d-1$ conditions, e.g., the condition of passing through $3d - 1$ general points.

Definition. Denote by N_d the number of rational plane curves of degree d that pass through $3d - 1$ given points in general position.

By arguments similar to those above, this number can also be characterized as the degree of the closure $\overline{V}_0^d \subset \mathbb{P}^{d(d+3)/2}$ (the most obvious compactification).

3.1.6 Example. For rational quartics, in the spirit of the example of the rational cubics, we must compute the degree of the subvariety $\overline{V}_0^4 \subset \mathbb{P}^{14}$ corresponding to quartics with three double points. This can still be done with classical methods. In fact, the number $N_4 = 620$ was determined by Zeuthen [87] in 1873. For rational quintics, one has to impose 6 (= genus) double points, and the number $N_5 = 87304$ was determined only in recent times. It was computed explicitly in [82]; previously, Ran [72] had indicated a recursion which determines, in principle, the number N_d for any d.

3.1.7 Severi varieties. The variety $\overline{V}_0^d$ is an example of a *Severi variety*. More generally one can study the varieties $V_g^d \subset \overline{V}_g^d \subset \mathbb{P}^{d(d+3)/2}$, consisting of all plane curves of degree d and genus g. (See for example Harris–Morrison [43].) The problem of determining the degree of $\overline{V}_g^d$ was solved only recently, see Caporaso–Harris [12] and Ran [72].

3.1.8 Parametrizations. The work of Kontsevich and Manin [58] has given us the crystal clear and explicit relation 3.3.1, which determines all the numbers N_d. Kontsevich's approach dramatically changes the point of view: instead of characterizing a curve by its equation (a point in a Severi variety), one studies its parametrization. The result is obtained as an intersection number in the moduli space $\overline{M}_{0,n}(\mathbb{P}^2, d)$ rather than $\overline{V}_0^d$ (as we shall see in the proof of Theorem 3.3.1). Note that $\overline{M}_{0,0}(\mathbb{P}^2, d)$ and $\overline{V}_0^d$ are birationally equivalent: both are compactifications of the open set V_0^d of irreducible and reduced rational curves.

A striking novelty of the approach is that the intersection numbers are computed without much knowledge of the intersection ring. Instead, the recursive structure of the moduli space is explored; as we shall see in the next section, the formula is a consequence of the fact that the boundary divisors of $\overline{M}_{0,n}(\mathbb{P}^r, d)$ are products of moduli spaces of lower dimension.

3.2 Counting conics and rational cubics via stable maps

To see how the recursion works, we will first recover the number of conics passing through five general points, and next, the number of rational cubics passing through eight points. The method reduces the question to the case of lower degree, and the starting point is simply this:

3.2.1 Fact. Through two distinct point there is a unique line. That is, $N_1 = 1$.

3.2.2 Proposition. *There is exactly $N_2 = 1$ conic passing through five general points in the plane.*

Proof. The computation takes place in $\overline{M}_{0,6}(\mathbb{P}^2, 2)$, a variety of dimension 11. Let us use the symbols $m_1, m_2, p_1, \ldots, p_4$ to indicate the marks. Take two lines L_1, L_2 in $\mathbb{P}^2$ and four points $Q_1, \ldots, Q_4$, all in general position. The content of the genericity assumption will be discussed along the way.

Let $Y \subset \overline{M}_{0,6}(\mathbb{P}^2, 2)$ be the subset consisting of maps

$$(C; m_1, m_2, p_1, \ldots, p_4; \mu) \text{ such that } \begin{cases} \mu(m_1) & \in L_1 \\ \mu(m_2) & \in L_2 \\ \mu(p_i) & = Q_i, \quad i = 1, \ldots, 4. \end{cases}$$

Y is in fact a subvariety, given by the intersection of the six inverse images under the evaluation maps:

$$Y = \nu_{m_1}^{-1}(L_1) \cap \nu_{m_2}^{-1}(L_2) \cap \nu_{p_1}^{-1}(Q_1) \cap \cdots \cap \nu_{p_4}^{-1}(Q_4).$$

By flatness of the evaluation maps, the inverse image of a line is of codimension 1, and the inverse image of a point is of codimension 2, so the total codimension of the intersection is 10. Choosing the lines and points in a sufficiently general way, one can ensure that Y is in fact of this codimension, in other words, Y is a curve. Now we are going to compute the intersection of Y with boundary divisors. The generality of the points and lines also imply (cf. 3.4.3) that Y intersects each boundary divisor transversally, and that the intersection takes place in $\overline{M}^*_{0,6}(\mathbb{P}^2, 2) \subset \overline{M}_{0,6}(\mathbb{P}^2, 2)$; this locus is smooth.

Consider the map $\overline{M}_{0,6}(\mathbb{P}^2, 2) \to \overline{M}_{0,\{m_1,m_2,p_1,p_2\}}$, which forgets the map μ as well as the two marks p_3, p_4. The fundamental linear equivalence 2.7.5.1 yields

$$Y \cap D(m_1, m_2|p_1, p_2) \equiv Y \cap D(m_1, p_1|m_2, p_2). \tag{3.2.2.1}$$

Let us first have a look at the special boundary divisor of the left-hand side, $D(m_1, m_2|p_1, p_2) = \sum D(A, B; d_A, d_B)$. There are 12 terms in this sum. The 12 irreducible components of the divisor correspond to the possible ways of distributing marks and degrees, as illustrated below. The twig with the A-marks is always drawn to the left. The numbers displayed close to the twigs are the partial degrees, d_A (to the left) and d_B (to the right):

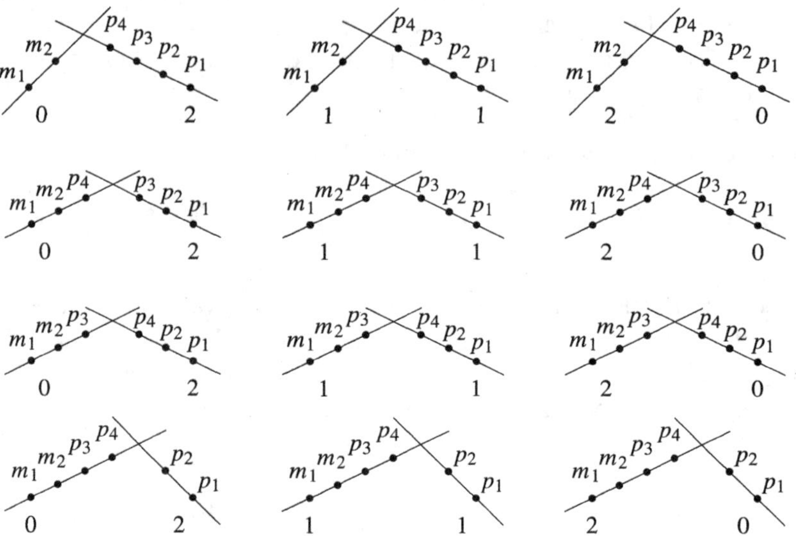

Let us compute the intersection of Y with each one of the irreducible boundary divisors. In the first column we have $d_A = 0$. This means that the curve C_A maps to a point $z \in \mathbb{P}^2$. Recalling that the marked point m_1 maps to L_1 and the marked point m_2 maps to L_2, we conclude that $\{z\} = L_1 \cap L_2$. Now suppose there were more marked points on C_A (in other words we are looking at the last three boundary divisors of the first column): then these spare marked points would also be mapped to z, in contradiction to the hypothesis of general position of the points and lines. This shows that Y has empty intersection with each of the last three divisors in the first column. As to the first case, we are then mapping C_B to a conic, and once this conic is fixed there are no more choices left for the marks, because the node $x \in C_A \cap C_B$ maps to z, and the other marked points map to the Q_i's. The number of ways to draw a conic through the five points is exactly N_2. So the sought-for number appears at this place in the sum.

In the third column, we have $d_B = 0$. This means that C_B maps to a point. But this is impossible because of the conditions defining Y: two of the marked points would map to the same point in $\mathbb{P}^2$. Thus there is no contribution at all from this column.

In the middle column, we have $d_A = d_B = 1$. Thus each twig is mapped to a line. In the first three cases, there are at least three marked points on C_B. They would all map to the same line, and the points would then be collinear, contradicting the generality requirement. Therefore, only the last configuration may give any contribution. Here C_A and C_B are mapped to distinct lines (otherwise all the points would be collinear). The image line $\mu(C_B)$ is uniquely determined since there is only $N_1 = 1$ line passing through Q_1 and Q_2, and similarly the image line $\mu(C_A)$ is determined uniquely by Q_3 and Q_4. Now let us count how many stable maps there are onto these lines, subject to the conditions. C_B has three special points: $p_1 \mapsto Q_1$, $p_2 \mapsto Q_2$, and finally the node x (where it is attached to C_A) must necessarily map to the unique point in $\mu(C_A) \cap \mu(C_B)$, so there is no choice for the position of the marked points on C_B. Similarly for C_A: note that the two marked points m_1 and m_2 are uniquely determined since they must map to the intersections $\mu(C_A) \cap L_1$ and $\mu(C_A) \cap L_2$, respectively. In conclusion, the intersection of Y with that divisor consists of a single point.

Summing up the contributions from all components of $D(m_1, m_2 | p_1, p_2)$, we get

$$Y \cap D(m_1, m_2 | p_1, p_2) = N_2 + 1.$$

Next we compute the intersection of Y with the divisor $D(m_1, p_1 | m_2, p_2)$. Again we could draw all the 12 components of this divisor, but let us do without. Since there is a p_j and an m_i on each twig, we cannot have any partial degree $d_k = 0$: otherwise this would force $Q_j \in L_i$, contradicting the generality of the L_j, Q_i. So we are left with the case $d_A = d_B = 1$. Here the only possibilities are p_3 on one twig and p_4 on the other. In each case the possibilities are reduced to that of drawing a line through two distinct points ($N_1 = 1$), yielding

$$Y \cap D(m_1, p_1 | m_2, p_2) = 1 + 1.$$

So in conclusion, the intersection of Y with the equivalence 3.2.2.1 yields $N_2 + 1 = 1 + 1$, and thus $N_2 = 1$. □

3.2.3 Proposition. *There are precisely $N_3 = 12$ rational cubics passing through 8 given points in general position.*

Proof. The line of argument is exactly the same as for the case of conics; only a little care is needed to determine the coefficients.

This time we place ourselves in $\overline{M}_{0,9}(\mathbb{P}^2, 3)$, a space of dimension 17. We denote the marks by $m_1, m_2, p_1, \ldots, p_7$, and consider the forgetful map to $\overline{M}_{0,4}$ which forgets the marks $p_3, \ldots, p_7$ (as well as the map). Fix two lines L_1, L_2 and seven points $Q_1, \ldots, Q_7$ in general position in $\mathbb{P}^2$. Let $Y \subset \overline{M}_{0,9}(\mathbb{P}^2, 3)$ be the curve defined as

$$Y = \nu_{m_1}^{-1}(L_1) \cap \nu_{m_2}^{-1}(L_2) \cap \nu_{p_1}^{-1}(Q_1) \cap \cdots \cap \nu_{p_7}^{-1}(Q_7).$$

One can show that Y intersects each of the boundary divisors transversally and is wholly contained in the locus $M_{0,9}^*(\mathbb{P}^2, 3)$, cf. 3.4.2.

The relation $Y \cap D(m_1, m_2|p_1, p_2) \equiv Y \cap D(m_1, p_1|m_2, p_2)$ will reveal an expression for N_3 in terms of N_2 and N_1.

Let us first compute the intersection of Y with $D(m_1, m_2|p_1, p_2)$. This divisor has 128 irreducible components! Indeed, there are five further marks to distribute on the two twigs; the number of ordered partitions $A \cup B = [5]$ is 32, which is then multiplied by the number 4 of partitions $d_A + d_B = 3$. As in the case of conics, let us examine each of the divisors $D(A, B; d_A, d_B)$ according to the partition $d_A + d_B = 3$.

If $d_B = 0$, the curve C_B is mapped to a single point. This is absurd, because it has marked points mapping to distinct points Q_i. Therefore Y has empty intersection with each of the boundary divisors with $d_B = 0$. If $d_A = 0$, then as in the case of conics, the entire twig C_A maps to the point $z \in L_1 \cap L_2$. We see that the choices of C_B correspond to the possible ways of drawing a rational cubic through the 8 points $z, Q_1, \ldots, Q_7$ (and once the image cubic is fixed, the position of the special points is determined by the requirements defining Y). Hence the term N_3 appears at this stage of the sum.

Let us consider the cases with $d_A = 1$. Unless we put precisely two extra marks on C_A and three extra marks on C_B, we get a contradiction with the generality assumption. Indeed, more than two spare marks on C_A would imply at least three collinear points on the image line $\mu(C_A)$; more than three spare marks on C_B would require at least six points on the conic image $\mu(C_B)$, which would also contradict the generality. Now there are $\binom{5}{2} = 10$ ways to distribute the remaining five marks, so we are dealing simultaneously with ten components; this gives a coefficient 10. For each of these components there is only $N_1 = 1$ choice for the image line $\mu(C_A)$ and $N_2 = 1$ choice for the image conic $\mu(C_B)$, so this determines each of the partial maps $C_A \to \mathbb{P}^2$ and $C_B \to \mathbb{P}^2$. There is no choice for the position of the marked points here, because the map is birational onto its image, and the marked points must be the inverse images of the given points Q_i (and the marked points m_1 and m_2 on C_A must be the unique inverse image of the intersections $\mu(C_A) \cap L_1$ and $\mu(C_A) \cap L_2$). It remains to describe how the two maps are glued together: there are two possibilities, namely the inverse images of the $2 = d_A \cdot d_B$ points of intersection $\mu(C_A) \cap \mu(C_B)$. Hence the contributions from the ten divisors with $d_A = 1$ amounts to

$$10 \cdot N_1 \cdot 2 \cdot N_2 = 20.$$

Now check the case $d_A = 2$. Arguing once again with the generality of points and lines, we conclude that only when all the five spare marks fall on C_A do we get any contribution. So we are now in the situation in which there is only one irreducible boundary component to consider. Again there is $N_2 = 1$ choice for the image conic $\mu(C_A)$ and $N_1 = 1$ choice for the image line $\mu(C_B)$. All the p-marks are determined uniquely by the requirements defining Y. But for the marked point m_1 on C_A there are two choices: it can be any one of the inverse image points of the intersection $L_1 \cap \mu(C_A)$. The same goes for m_2. This accounts for a factor 2^2. Finally there is a factor $2 = d_A \cdot d_B$ for the choices of where to glue the two twigs (just as above), giving a total coefficient equal to 8.

Grand total:

$$Y \cap D(m_1, m_2 | p_1, p_2) = N_3 + 20 + 8.$$

Now let us turn our attention to the points of intersection of Y with the special boundary divisor $D(m_1, p_1 | m_2, p_2)$. Since there is both an m and a p on each twig, there is no contribution from the cases with $d_A = 0$ or $d_B = 0$. For $d_A = 1$ there must be exactly one more marked point on C_A. There are five ways to choose this mark among the remaining marks $p_3, \dots, p_7$. So here we are considering five irreducible components in one go. The two image curves are now determined: C_A is the line through the two points, and C_B is the unique conic through 5 points. The p-marks are uniquely determined; for m_1 (on C_A) there is $d_A = 1$ possibility, and for m_2 (on C_B) there are $d_2 = 2$ ways. There are two ways of gluing the two partial maps, corresponding to the $d_A \cdot d_B = 2$ intersection points of the image curves. Total: $5 \cdot 2 \cdot 2 = 20$.

The situation is symmetric when $d_A = 2$ since then $d_B = 1$. This accounts for another 20 maps, giving a total of $Y \cap D(m_1, p_1 | m_2, p_2) = 20 + 20$.

Finally, since the two special boundary divisors are equivalent, we can write

$$N_3 + 20 + 8 = 20 + 20,$$

whence $N_3 = 12$ as claimed. □

3.3 Kontsevich's formula for rational plane curves

3.3.1 Theorem. (Kontsevich) *Let N_d be the number of rational curves of degree d passing through $3d - 1$ general points in the plane. Then the following recursive relation holds:*

$$N_d + \sum_{\substack{d_A + d_B = d \\ d_A \geq 1, d_B \geq 1}} \binom{3d-4}{3d_A - 1} d_A^2 N_{d_A} \cdot N_{d_B} \cdot d_A d_B = \sum_{\substack{d_A + d_B = d \\ d_A \geq 1, d_B \geq 1}} \binom{3d-4}{3d_A - 2} d_A N_{d_A} \cdot d_B N_{d_B} \cdot d_A d_B.$$

Since we know $N_1 = 1$, the formula allows for the computation of any N_d.

Proof. Set $n := 3d$. Consider $\overline{M}_{0,n}(\mathbb{P}^2, d)$, with marks named $m_1, m_2, p_1, \dots, p_{n-2}$. Let L_1 and L_2 be lines and let $Q_1, \dots, Q_{n-2}$ be points in $\mathbb{P}^2$. Let $Y \subset \overline{M}_{0,n}(\mathbb{P}^2, d)$ be defined as the intersection of the inverse images of these points and lines under the evaluation maps. The points and lines can be chosen in such a way that Y is a curve intersecting the boundary transversally and is wholly contained in the locus $M^*_{0,n}(\mathbb{P}^2, d)$ (cf. 3.4.2).

As in the cases $d = 2, 3$ considered above, the result will follow from the fundamental equivalence

$$Y \cap D(m_1, m_2|p_1, p_2) \equiv Y \cap D(m_1, p_1|m_2, p_2).$$

Let us examine the left-hand side. The only contribution with a partial degree equal to zero comes from the case in which all the $3d - 4$ spare marks fall on the B-twig, and the number of ways to draw the corresponding curve is by definition N_d. When the partial degrees are positive, the only distributions of the marks giving contribution is when $3d_A - 1$ marks fall on the A-twig. There are $\binom{3d-4}{3d_A-1}$ such irreducible components in $D(m_1, m_2|p_1, p_2)$, thus explaining this binomial factor in the formula. Now there are N_{d_A} ways to draw the image of C_A, and N_{d_B} ways to draw the image of C_B, and then the position of all the p_i's is determined. It remains to choose where to put the two marked points m_1, m_2. The marked point m_1 has to fall on a point of the intersection of $\mu(C_A)$ with L_1, and by Bézout's theorem there are d_A such points; same thing for m_2. This accounts for the factor d_A^2 in the formula. Finally, the intersection point $x \in C_A \cap C_B$ must go to one of the $d_A \cdot d_B$ points of intersection of the two image curves (Bézout again). This explains the factor $d_A d_B$ and completes the examination of the left-hand side of the equation.

On the right-hand side, we get no contribution when d_A or d_B is zero: this would imply $Q_1 \in L_1$ or $Q_2 \in L_2$, arguing as in the two examples above. For the other possible partitions $d_A + d_B = d$, the only contribution comes from components with $3d_A - 2$ further marks on the A-twig, and there are $\binom{3d-4}{3d_A-2}$ such components. For each of these components, the image curves $\mu(C_A)$ and $\mu(C_B)$ can be chosen in N_{d_A} and N_{d_B} ways, respectively. The marked point m_1 must map to $\mu(C_A) \cap L_1$, giving d_A choices, and similarly m_2 allows d_B choices. Finally, to glue these two maps, there is the choice among any one of the $d_A d_B$ points of intersection $\mu(C_A) \cap \mu(C_B)$. This completes the proof. □

3.4 Transversality and enumerative significance

In this section we establish the transversality results used in the proof of Kontsevich's formula and in the two preceding examples. We also check that counting stable maps is the same as counting curves!

3.4.1 Notation. Let us start out introducing some shorthand notation. We set $\overline{M} := \overline{M}_{0,n}(\mathbb{P}^r, d)$ and let $\{p_1, \ldots, p_n\}$ denote the set of marks. Set $X := \mathbb{P}^r$. Let $X^n = X \times \cdots \times X$ be the product of n factors equal to X and let $\tau_i : X^n \to X$ be the ith projection. Given n irreducible subvarieties $\Gamma_1, \ldots, \Gamma_n \subset X$, let $\underline{\Gamma}$ denote their product:

$$\underline{\Gamma} := \Gamma_1 \times \cdots \times \Gamma_n = \bigcap \tau_i^{-1}(\Gamma_i) \subseteq X^n.$$

The n evaluation maps $\nu_i : \overline{M} \to X$ induce a map $\underline{\nu} : \overline{M} \to X^n$. In other words, for each $i = 1, \ldots, n$, we have a commutative diagram

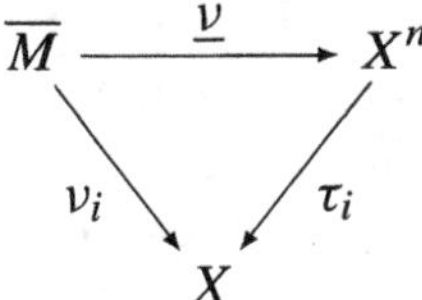

The inverse image $\nu_i^{-1}(\Gamma_i) \subset \overline{M}$ consists of all maps μ such that $\mu(p_i) \in \Gamma_i$. If k_i is the codimension of Γ_i in $\mathbb{P}^r$, then $\nu_i^{-1}(\Gamma_i)$ is of the same codimension k_i in $\overline{M}$ (by flatness 2.5.1). The intersection (as schemes)

$$\nu_1^{-1}(\Gamma_1) \cap \cdots \cap \nu_n^{-1}(\Gamma_n) = \underline{\nu}^{-1}(\underline{\Gamma})$$

is the locus of maps μ such that $\mu(p_i) \in \Gamma_i$, for $i = 1, \ldots, n$. In particular, the image of each of these maps μ meets each Γ_i. Note that since $\underline{\nu}$ is not flat (cf. 2.5.3), this locus is not automatically of the expected codimension $\sum k_i$.

We are mostly interested in the situation $\sum \operatorname{codim} \Gamma_i = \dim \overline{M}$. In that case we would expect the intersection of the inverse images to be of dimension zero, so that only finitely many maps satisfy the conditions. The proposition below asserts that this is indeed the case in the generic situation.

Let us first recall the theorem of Kleiman on the transversality of the general translate (cf. [49]). Let G be a connected algebraic group. Let X be an irreducible variety with a transitive G-action; let $f : Y \to X$ and $Z \to X$ be morphisms between irreducible varieties. For each $\sigma \in G$, denote by Y^σ the variety Y considered as a variety over X via the composition $\sigma \circ f$.

3.4.2 Theorem. (Kleiman [49]) *There exists a dense open subset $U \subset G$ such that for every $\sigma \in U$, the fiber product $Y^\sigma \times_X Z$ is either empty or we have*

$$\dim(Y^\upsilon \times_X Z) = \dim Y + \dim Z - \dim X.$$

Furthermore, if Y and Z are smooth, then U can be chosen such that for every $\sigma \in U$, the fiber product $Y^\sigma \times_X Z$ is also smooth. □

3.4.3 Proposition. *For generic choices of $\Gamma_1, \ldots, \Gamma_n \subset \mathbb{P}^r$, with codimensions adding up to* $\dim \overline{M}$*, the scheme-theoretic intersection*

$$\underline{\nu}^{-1}(\underline{\Gamma}) = \bigcap_{i=1}^{n} \nu_i^{-1}(\Gamma_i)$$

consists of a finite number of reduced points, supported in any preassigned nonempty open set, and in particular, in the locus $M^* \subset \overline{M}$ of maps with smooth source and without automorphisms.

Proof. By abuse of notation we will also write M^* for the given nonempty open set. Let $\underline{G}$ denote the product of n copies of the group G. It acts transitively on X^n. Repeated use of Kleiman's theorem will imply the proposition.

First we apply the theorem to the complement $(M^*)^{\complement}$; this is a closed subvariety of codimension at least 1 in $\overline{M}_{0,n}(\mathbb{P}^r, d)$. The inverse image in $(M^*)^{\complement}$ of a translate $\underline{\Gamma}^\sigma$ is identified with the fiber product $\underline{\Gamma}^\sigma \times_{X^n} (M^*)^{\complement}$. Kleiman's theorem applied to

$$\begin{array}{ccc} & & (M^*)^{\complement} \\ & & \big\downarrow \underline{\nu} \\ \underline{\Gamma} & \hookrightarrow & X^n \end{array}$$

gives us a dense open set $V_1 \subset \underline{G}$ such that the inverse image in $(M^*)^{\complement}$ of any of the translates $\underline{\Gamma}^\sigma$, with $\sigma \in V_1$, is empty. Therefore, in general the intersection is wholly supported in M^* as asserted.

Now apply Kleiman to

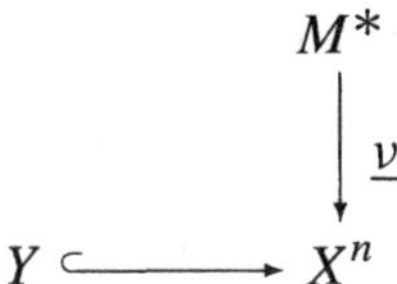

with $Y := \operatorname{Sing} \underline{\Gamma}$. We obtain a dense open set $V_2 \subset \underline{G}$ such that $\underline{\nu}^{-1}(Y^\sigma) = \emptyset$. Now let $Y = \underline{\Gamma} \setminus \operatorname{Sing} \underline{\Gamma}$. Since the varieties in the diagram are now smooth, we find a dense open set $V_3 \subset \underline{G}$ such that the inverse image in M^* of each of the corresponding translates is of correct dimension (or is empty), and also is smooth. Hence it consists of a finite number of reduced points (possibly zero).

Consequently, for all the translates under $\sigma \in V_1 \cap V_2 \cap V_3 \subset \underline{G}$, the corresponding inverse image is of the correct dimension, is reduced, and is supported in the given open set. □

3.5 Stable maps versus rational curves

3.5.1 What are we counting? Maps were not the type of objects we intended to count in the first place. We were really interested in counting rational curves, without mentioning either marks or maps. Now since each solution map sends p_i to Γ_i, then in particular the image curve meets each Γ_i. That is, we have here all the solutions to the question "how many rational curves meet the Γ_i's?". It remains to check whether there is any repetition, that is, if any of the solution curves intersects any Γ_i in more than one point. In this case, this single rational curve would correspond to two or more n-pointed stable maps satisfying the conditions $\mu(p_i) \in \Gamma_i$, due to the different ways of putting the marks on the same curve.

If any Γ_i is a hypersurface, this type of repetition is unavoidable. Indeed, if $\Gamma_i \subset \mathbb{P}^r$ is a hypersurface of degree e_i, then by Bézout's theorem, a curve of degree d will *always* meet Γ_i; the number of points (if finite) is $d \cdot e_i$, counted with multiplicity. So for each rational curve that is a solution to the question of incidences, there are $\prod_i d \cdot e_i$ nonisomorphic stable n-pointed maps satisfying the corresponding condition. We must preclude this case, or make the necessary correction by the factor $d \cdot e_i$ for each hypersurface Γ_i, as we will do in Lemma 4.2.4 below.

If codim $\Gamma_i \geq 2$, then it is most likely that the curve does not intersect Γ_i at all. However, since we are forcing it to do so, it will most likely intersect at just one point. In other words, since we are demanding the curve to do more than its codimension naturally makes us expect, it should not, by its own initiative, meet Γ_i at more than one point. So if all the varieties Γ_i are of codimension at least 2, it is expected that all the solution maps meet each Γ_i at just one point. In this case, the number of solutions to the problem of counting n-pointed stable maps is equal to the solution of the problem for rational curves (without mention of marks).

The rest of this section is dedicated to the formalization of this discussion. There are two types of behavior we want to exclude: the first is the situation in which the same curve passes twice through the same point, and the second is when the curve meets Γ_i at two or more distinct points.

3.5.2 Lemma. *Suppose $n \geq 2$. Consider the locus*

$$Q_{ij} := \{\mu \in M_{0,n}(\mathbb{P}^r, d) \mid \mu(p_i) = \mu(p_j)\}$$

of maps whose two marked points $p_i \neq p_j$ have common image in $\mathbb{P}^r$. Then the codimension of Q_{ij} in $M := M_{0,n}(\mathbb{P}^r, d)$ is equal to r.

Note that here we are talking about M and not $\overline{M}$. This is enough since we have already excluded the possibility that there could be any reducible solutions.

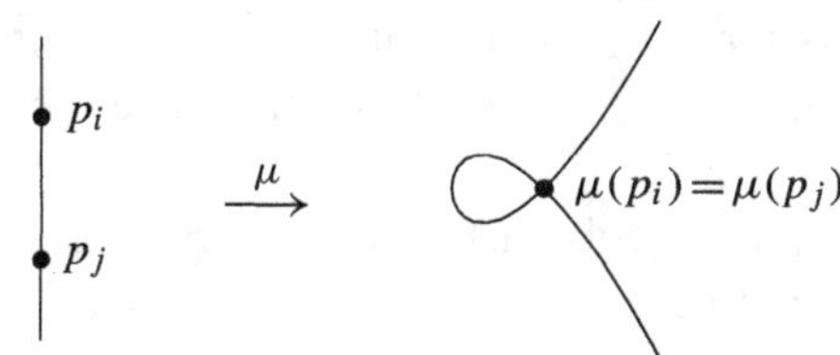

Proof. We can assume $n \geq 3$ by a reduction similar to that of 2.8.3, so we can work in the space $W(r, d)$ of $(r+1)$-tuples of degree-d forms (see 2.1.1). Let $a_{k0}x^d + a_{k1}x^{d-1}y + \cdots + a_{kd}y^d$ be the kth form. Assuming $p_i = [0 : 1]$ and $p_j = [1 : 0]$, the condition $\mu(p_i) = \mu(p_j)$ reads

$$[a_{00}, a_{10}, \ldots, a_{r0}] = \lambda[a_{0d}, a_{1d}, \ldots, a_{rd}]$$

for some $\lambda \in \mathbb{C}^*$, which makes up r independent conditions in the a_{ij} (cf. the argument of 2.1.2). Alternatively, the codimension of the set of zeros of the 2×2 minors of the matrix $(a_{ij})_{0 \leq i \leq r, j=0,d}$ is r. $\square$

3.5.3 Lemma. *For generic choices of* $\Gamma_1, \ldots, \Gamma_n \subset \mathbb{P}^r$*, with codimensions adding up to* $\dim \overline{M}_{0,n}(\mathbb{P}^r, d)$*, we have*

$$\mu^{-1}\mu(p_i) = \{p_i\},\ i = 1, \ldots, n \text{ (with multiplicity 1)}$$

for every map μ *in the intersection* $\underline{\nu}^{-1}(\underline{\Gamma})$.

This means that generically we have

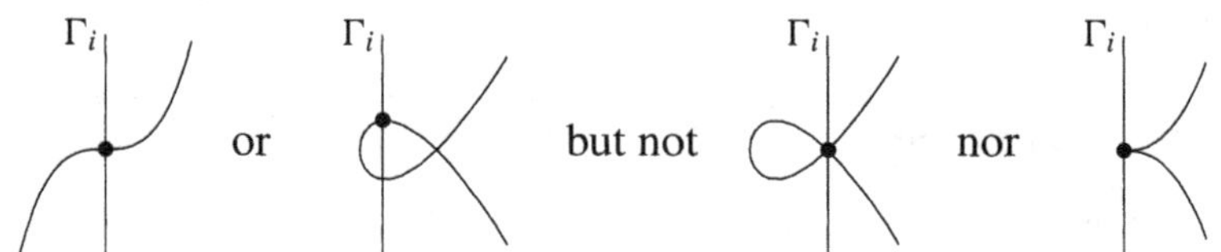

Proof. By Kleiman's theorem, for general translates of Γ_i, the intersection $\underline{\nu}^{-1}(\underline{\Gamma})$ consists of a finite number of reduced points, supported in the dense open set $M^\circ_{0,n}(\mathbb{P}^r, d)$ of immersions with smooth source (cf. 2.1.2). This already shows that $\mu^{-1}\mu(p_i)$ is reduced for each $i = 1, \ldots, n$. Now inside $M^\circ_{0,n}(\mathbb{P}^r, d)$ we have to avoid (for each i) the locus J_i of maps μ for which the preimage of $\mu(p_i)$ contains at least one point distinct from p_i. If we show that this locus is of positive codimension, the result clearly follows from yet another transversality argument.

Step up to the space $M^\circ_{0,n+1}(\mathbb{P}^r, d)$ with one extra mark named p_0, and consider the forgetful map $\varepsilon : M^\circ_{0,n+1}(\mathbb{P}^r, d) \to M^\circ_{0,n}(\mathbb{P}^r, d)$, which forgets p_0. We claim that the image of $Q_{i,0}$ is exactly $J_i \subset M^\circ_{0,n}(\mathbb{P}^r, d)$. Indeed, it is clear that the image is contained in J_i. On the other hand ε is surjective. In fact, for each map $\mu \in J_i$ we know there exists a point, other than p_i, with the same image. So putting the extra mark at this point we get an $(n+1)$-pointed map belonging to $Q_{i,0}$ and whose image is μ.

Finally, since $Q_{i,0}$ has codimension r, we conclude that J_i has codimension at least $r - 1 \geq 1$, as claimed. $\square$

3.5.4 Corollary. *If $\Gamma_1, \dots, \Gamma_{3d-1}$ are general points in $\mathbb{P}^2$, then the number of stable maps such that $p_i \mapsto \Gamma_i$ is equal to the number N_d of rational curves through those points.*

Proof. By Lemma 3.5.3, each solution map passes only once through each point, so there is precisely one possibility for the position of each marked point; hence the number is also the number of rational curves passing through the points, without mention of marks. □

In higher-dimensional projective spaces, there are other interesting subvarieties than points to impose incidence to. For example, in $\mathbb{P}^3$ it is natural to impose the condition of being incident to a given line; cf. also Example 3.1.1. In this case there is yet another case we must exclude in order to be sure that counting maps is the same as counting their image curves, namely the possibility of having a map that passes several times through the same Γ_i, but at distinct points.

3.5.5 Lemma. *Let $\Gamma_1, \dots, \Gamma_n \subset \mathbb{P}^r$ be general subvarieties of codimension at least 2, and with codimensions adding up to* $\dim \overline{M}_{0,n}(\mathbb{P}^r, d)$. *Then, for any $\mu \in \underline{\nu}^{-1}(\underline{\Gamma})$, the image curve $\mu(C)$ intersects each Γ_i at just one single point ($\mu(p_i)$).*

In other words, generically, when the codimension of Γ_i is at least two, the lemma allows (in principle)

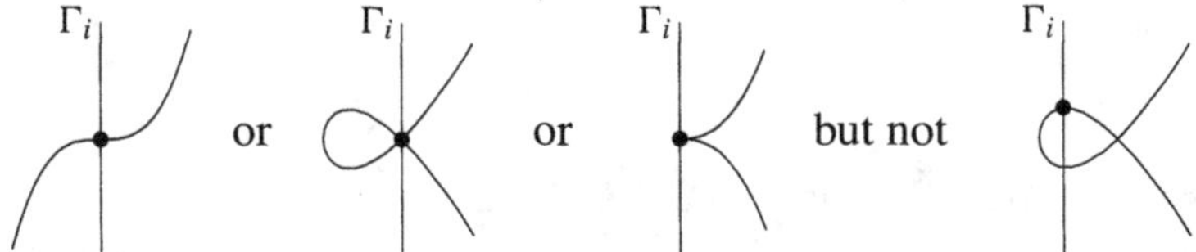

and together with Lemma 3.5.3, it follows that only the first figure above survives.

Proof. Let us work with the first mark, and afterward repeat the argument with the remaining marks. Since we have already excluded the possibility that reducible solution maps could occur, it is enough to work in $M_{0,n}(\mathbb{P}^r, d)$. Step up to the space $M_{0,n+1}(\mathbb{P}^r, d)$ with one extra mark p_0, and consider here the open set $M^\# := M_{0,n+1}(\mathbb{P}^r, d) \smallsetminus Q_{1,0}$ of maps with $\mu(p_1) \neq \mu(p_0)$. We will show that for generic choices of the Γ_i's, the intersection $\nu_0^{-1}(\Gamma_1) \cap \underline{\nu}^{-1}(\underline{\Gamma}) \cap M^\#$ is empty.

We keep the notation $X := \mathbb{P}^r$, $\underline{G} := \mathrm{Aut}(X)^n$. Consider the action of $\underline{G}$ on X^{n+1} (one extra factor) defined by

$$(g_i) \cdot (x_0, x_1, \dots, x_n) = (g_0 \cdot x_0, g_1 \cdot x_1, g_2 \cdot x_2, \dots, g_n \cdot x_n),$$

where $g_i \in \mathrm{Aut}(X)$ $x_i \in X$, and we take as the extra factor $g_0 := g_1$. Restricting to the complement $U_{01} \subset X^{n+1}$ of the diagonal $x_0 = x_1$, we get a transitive action.

Set $\Gamma_0 := \Gamma_1$ and consider the $n+1$ evaluation maps $\underline{\widetilde{\nu}} : M^\# \to U_{01}$. Look at the intersection

$$M^\# \cap \nu_0^{-1}(\Gamma_0) \cap \underline{\nu}^{-1}(\underline{\Gamma}) = \underline{\widetilde{\nu}}^{-1}(\Gamma_0 \times \underline{\Gamma})$$

inside $M^\#$. Note that the codimension of $\Gamma_0 \times \underline{\Gamma}$ in $X \times X^n$ is equal to codim $\Gamma_1 +$ dim $\overline{M}_{0,n}(\mathbb{P}^r, d) >$ dim $M^\#$ by the assumption. Arguing as in the proof of Lemma 3.4.3, we conclude via Kleiman that this intersection in $M^\#$ is empty for generic choices of Γ_i. More precisely, there exists a dense open set of $\underline{G}$ constituted by (g_i)'s such that

$$\nu_0^{-1}(g_0 \cdot \Gamma_0) \cap \underline{\nu}^{-1}(g \cdot \underline{\Gamma}) \cap M^\# = \emptyset.$$

Since we have already shown that the conditions have empty intersection with $Q_{1,0}$ and with the boundary, in fact $\nu_0^{-1}(g_0 \cdot \Gamma_0) \cap \underline{\nu}^{-1}(g \cdot \underline{\Gamma})$ has empty intersection with the whole of $\overline{M}_{0,n+1}(\mathbb{P}^r, d)$.

Now let us go back to the original space to complete the argument. Consider in $\overline{M}_{0,n}(\mathbb{P}^r, d)$ the intersection $\underline{\nu}^{-1}(g \cdot \underline{\Gamma})$ (for some g in the open set specified above) and suppose there exists herein a map μ that intersects Γ_1 at another point q, distinct from $\mu(p_1)$. Then putting the extra mark p_0 in the preimage $\mu^{-1}(q)$ (and stabilizing if necessary) we would get also an element of $\overline{M}_{0,n+1}(\mathbb{P}^r, d)$ in the intersection $\nu_0^{-1}(g_0 \cdot \Gamma_1) \cap \underline{\nu}^{-1}(g \cdot \underline{\Gamma}) \cap \overline{M}_{0,n+1}(\mathbb{P}^r, d)$, contradiction.

Repeating the argument with the other marks $p_2, \ldots, p_n$, we obtain the promised dense open set in $\underline{G}$. □

3.6 Generalizations and references

3.6.1 Ad hoc arguments for other targets. Even though it may be feasible to apply ad hoc arguments similar to those of this chapter to the case of low-degree rational curves in $\mathbb{P}^3$ or in other convex varieties, this is not recommended. The techniques of quantum cohomology described in the next two chapters provide a considerable simplification, computationally as well as conceptually. In the case of $\mathbb{P}^1 \times \mathbb{P}^1$, however, everything works out just as smoothly as in the case of $\mathbb{P}^2$. In the exercises to this chapter, we work out the analogue of Kontsevich's formula for $\mathbb{P}^1 \times \mathbb{P}^1$.

3.6.2 Tangency conditions and characteristic numbers. Asking that a plane curve be tangent to a given line $L \subset \mathbb{P}^2$ is a condition of codimension 1, i.e., it defines a divisor in $\overline{M}_{0,n}(\mathbb{P}^2, d)$ (or in $\overline{V}_0^d$ (3.1.5)). The *characteristic numbers* of a system of plane curves are defined as the number of curves passing through a points and tangent to b lines. If the system is the family of rational curves of degree d, we must have $a + b = 3d - 1$ for the question to make sense.

The characteristic numbers for plane curves of degree $d = 2, 3, 4$ were computed in the 19th century by Chasles, Maillard, and Zeuthen, respectively, and the verification of their results has been a challenge for modern enumerative geometry. Many of these numbers were verified with rigor in the eighties, using various ingenious compactifications of the open Severi varieties.

The advent of the Kontsevich moduli spaces has advanced the subject tremendously. For rational curves, the problem was solved by Pandharipande in [69]: he computes the class of the tangency divisor and gives an algorithm that permits the determination of all the genus-0 characteristic numbers, for any degree. The key step of the algorithm is the recursive structure of the boundary. A more powerful machinery was developed in Graber–Kock–Pandharipande [38]. The approach there is to use pointed conditions (i.e., require the tangency to occur at a given marked point of the map; this is a codimension-2 condition) and interpret the conditions in terms of certain tautological classes on the Kontsevich moduli space (cf. 4.5.5). This leads to concise formulas—also for genus 1 (and 2).

3.6.3 Genus 1. There is also a recursive formula (due to Eguchi–Hori–Xiong [21]) for the numbers E_d of plane curves of genus 1 and degree d passing through $3d$ general points, given in terms of E_d for lower degree, and the numbers N_d. (cf. for example Pandharipande [68]). Curiously, although this relation looks like it were a consequence of an equivalence of boundary divisors (just like Kontsevich's formula), no direct geometric interpretation is known.

Starting from this recursion, Vakil [84] extended the ideas of Pandharipande [69] to determine also the characteristic numbers for plane curves of genus 1. He identifies the good component of $\overline{M}_{1,0}(\mathbb{P}^2, d)$, describes its boundary, and gives a recipe to reduce questions of tangency to those of incidence, whose solutions E_d are known.

3.6.4 Plane quartics. Let us finally mention that Vakil [83] has verified the characteristic numbers of smooth plane quartics ($g = 3$), determined originally by Zeuthen [87]. The analysis takes place on the normalization of the good component of $\overline{M}_{3,0}(\mathbb{P}^2, 4)$.

Exercises

The quadric surface. The exercises to this chapter concern the quadric surface $Q \subset \mathbb{P}^3$ (given by the equation $X_0X_3 - X_1X_2 = 0$), and culminate with Kontsevich's formula for rational curves in Q.

Recall that Q is the image of the the Segre embedding

$$\begin{aligned} s : \mathbb{P}^1 \times \mathbb{P}^1 &\longrightarrow \mathbb{P}^3 \\ \left(\begin{bmatrix} x_0 \\ x_1 \end{bmatrix}, \begin{bmatrix} y_0 \\ y_1 \end{bmatrix} \right) &\longmapsto \begin{bmatrix} x_0 y_0 \\ x_0 y_1 \\ x_1 y_0 \\ x_1 y_1 \end{bmatrix}. \end{aligned}$$

The family of lines $s(\mathbb{P}^1, \left[\begin{smallmatrix} y_0 \\ y_1 \end{smallmatrix}\right])_{\left[\begin{smallmatrix} y_0 \\ y_1 \end{smallmatrix}\right] \in \mathbb{P}^1}$ are called horizontal rules; vertical rules are defined likewise. A curve in Q is said to be of bi-degree (m, n) if it is of class m times the class of a horizontal rule plus n times the class of a vertical rule. Bézout's theorem for Q says that two curves of bi-degrees (m, n) and (m', n') with no common component intersect in $mn' + m'n$ points (counted with appropriate multiplicities).

Clearly we have the following basic fact: through any point there is precisely one $(1, 0)$-curve and one $(0, 1)$-curve (i.e., precisely one horizontal rule and one vertical rule).

1. Count parameters to show that to get a finite number of rational (m, n)-curves, you need to impose $2m + 2n - 1$ point conditions.

   Let $N_{(m,n)}$ denote the number of rational curves of bi-degree (m, n) passing through $2m + 2n - 1$ general points on Q.

2. Show that $N_{(m,n)} = N_{(n,m)}$.

3. Imitate the argument in Proposition 3.2.2 to show that there is precisely

   $$N_{(1,1)} = 1$$

   $(1, 1)$-curve through 3 general points. (This is easy to see directly: we are talking about hyperplane sections of $Q \subset \mathbb{P}^3$, and three points span a plane.)

4. Let $C \subset Q$ be a curve of bi-degree $(1, 2)$ or $(2, 1)$. Show that C is a twisted cubic in $\mathbb{P}^3$, and in particular a rational curve. (All twisted cubics contained in Q are thus obtained.)

5. Assuming the result of the previous exercises, imitate the arguments in Proposition 3.2.2 and Proposition 3.2.3 to show that there is precisely

   $$N_{(2,1)} = 1$$

   $(2, 1)$-curve through 5 general points.

6. Generalizing the previous two exercises, show that $N_{(d,1)} = 1$ for all d. *Hint:* multiple covers of rules do not pass through many general points.

7. Assuming the result of the previous exercises, imitate the arguments in Proposition 3.2.3 to show that there are precisely

$$N_{(2,2)} = 12$$

rational (2, 2)-curves passing through 7 general points.

8. (Classical proof of $N_{(2,2)} = 12$.) Given 7 points on Q, there is a linear pencil of (2, 2)-curves through them: these are intersections of Q with a pencil of quadrics $\{t_1 F_1 + t_2 F_2\}_{[t_1:t_2]\in\mathbb{P}^1}$. This family has 8 base points, namely the points in the intersection $Q \cap F_1 \cap F_2$. Blow up these 8 base points to realize a family of (2, 2)-curves with base $\mathbb{P}^1$. The general member of the pencil is of genus 1, the special fibers are rational (2, 2)-curves with a single double point.

   Now proceed as in Example 3.1.4 to show that $N_{(2,2)} = 12$. (The surface Q has Euler characteristic 4.)

9. Imitate the proof of Theorem 3.3.1 to prove the $\mathbb{P}^1 \times \mathbb{P}^1$ analogue of Kontsevich's formula:

$$\begin{aligned} N_{(d,e)} &+ \sum_{\substack{d_A+d_B=d\\ e_A+e_B=e}} \binom{2d+2e-4}{2d_A+2e_A-1} \cdot d_A e_A N_{(d_A,e_A)} \cdot N_{(d_B,e_B)} \cdot (d_A e_B + d_B e_A) \\ &= \sum_{\substack{d_A+d_B=d\\ e_A+e_B=e}} \binom{2d+2e-4}{2d_A+2e_A-2} \cdot d_A N_{(d_A,e_A)} \cdot e_B N_{(d_B,e_B)} \cdot (d_A e_B + d_B e_A). \end{aligned}$$

   This result is also due to Kontsevich–Manin [58]. However, their formula (5.19) has a misprint in the last binomial coefficient.

Chapter 4

Gromov–Witten Invariants

The intersection numbers resulting from an ideal transverse situation as in Proposition 3.4.3 are the (genus-0) *Gromov–Witten invariants*. In Section 4.2 we establish the basic properties of Gromov–Witten invariants, and in 4.3 and 4.4 we describe recursive relations among them, allowing for their computation.

For simplicity, throughout this chapter we assume $r \geq 2$. See the exercises for a few comments on the case of $\mathbb{P}^1$.

4.1 Definition and enumerative interpretation

From the viewpoint of Chapter 3, the goal is to compute the number of points in the finite set $\underline{\nu}^{-1}(\underline{\Gamma})$, that is, to compute the degree $\int[\underline{\nu}^{-1}(\underline{\Gamma})]$. The problem here is that we are on a singular variety, and intersection of cycle classes may not be well defined. The product of operational classes, i.e., cohomology classes, is the right tool to make things work properly.

4.1.1 The cohomology ring of $\mathbb{P}^r$. For $X = \mathbb{P}^r$, indeed for any smooth variety, the Chow group $A_*(X)$ of cycle classes modulo rational equivalence is in fact a ring ([28], Ch. 6). The intersection ring $A^*(X)$ is defined by setting $A^k(X) := A_{r-k}(X)$ where $r = \dim X$. The isomorphism is the Poincaré duality isomorphism

$$\begin{array}{ccc} A^*(X) & \stackrel{\sim}{\to} & A_*(X) \\ \gamma & \mapsto & \gamma \cap [X]. \end{array}$$

For this reason, there is not much need to distinguish between the operational classes in $A^*(X)$ and cycle classes, and we will allow ourselves sometimes to use the notation $[\Gamma]$ also for the operational class corresponding to $[\Gamma] \in A_*(X)$ under Poincaré duality. Throughout we will work with $\mathbb{Q}$-coefficients. It so happens for $X = \mathbb{P}^r$ that the intersection ring $A^*(X)$ is isomorphic to the cohomology ring $H^*(X)$ (e.g., de Rham cohomology or singular cohomology), albeit with a doubling of degrees, $A^k(X) \simeq H^{2k}(X)$. For general smooth X, the cohomology ring is better behaved (e.g., satisfies the Künneth formula (cf. 4.3.1)), so we will refer to $A^*(X)$ as the cohomology ring.

4.1.2 Cohomology classes on $\overline{M}_{0,n}(\mathbb{P}^r, d)$. The moduli space $\overline{M} = \overline{M}_{0,n}(\mathbb{P}^r, d)$, however, is a singular variety so here there is no well-defined intersection product on the level of cycle classes. But there are cohomology classes that can be manipulated with the same ease as in the smooth situation. In fact, if $f : Y \to X$ is a map of an arbitrary scheme to a smooth variety, there is a well-defined product $A^k(X) \otimes A_i(Y) \to A_{i-k}(Y)$ (cf. [28], Ch. 17). Briefly, this can be seen as follows. Smoothness of X allows us to employ the diagonal construction. Let $V \subset X$ be a subvariety of codimension k and assume $Z \subset Y$ is integral of dimension i. In the diagram with Cartesian square below,

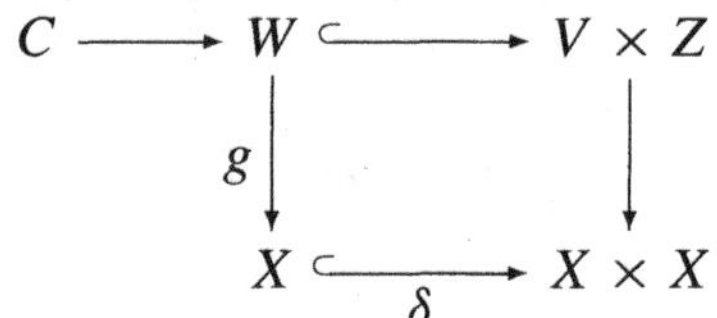

C denotes the normal cone of W in $V \times Z$. The normal cone sits as a closed subscheme of g^*N_δ, the pullback to W of the normal bundle of δ. It is of pure dimension $\dim(V \times Z) = n - k + i$, with $n = \dim X = \operatorname{rk} N_\delta$. Any cycle class in $A_r(g^*N_\delta)$ is the pullback of a unique cycle class in $A_{r-n}(W)$. Since we have $W = f^{-1}(V) \cap Z \subset Y$, the class of C produces the desired class in $A_{i-k}(Y)$ by proper pushforward. In this way we get lots of interesting operators on the Chow group $A_*(Y)$. Another source of useful operators are of course Chern classes. The subring $A^*(Y)$ of $\operatorname{End}(A_*(Y))$ spanned by these operators will play the role of a cohomology ring, and we will just call the operators *cohomology classes*. Just like Chern classes, they are operators you apply to cycle classes to obtain new cycle classes; the evaluation of an operator $\alpha \in A^*(Y)$ on a cycle $[Z] \in A_*(Y)$ is written $\alpha \cap [Z]$. The multiplication in $A^*(Y)$ is denoted $\cup$: given $\alpha, \beta \in A^*(Y)$ we define $\alpha \cup \beta$ by the rule $(\alpha \cup \beta) \cap [Z] = \alpha \cap (\beta \cap [Z])$ for any $[Z] \in A_*(Y)$. By abuse of language we shall say that a class α is of codimension k if $\alpha \in A^k(Y)$, i.e., $\alpha \cap [Z]$ lands in $A_{i-k}(Y)$ for any $[Z] \in A_i(Y)$; in particular, $\alpha \cap [Y]$ is a cycle class of codimension k. If $f : Y' \to Y$ is a morphism, we have the operation of pullback $f^* : A^*(Y) \to A^*(Y')$, whose properties are analogous to the properties of pullback of Chern classes, e.g., the projection formula holds.

The cohomology classes on $\overline{M}$ that we shall be concerned with are those pulled back from $\mathbb{P}^r$ via the evaluation maps. Let us fix some notation. Henceforth we put

$X = \mathbb{P}^r$. Let $\gamma_i \in A^*(X)$ be the cohomology class corresponding to $[\Gamma_i] \in A_*(X)$ via Poincaré duality. Then $\underline{\gamma} := \gamma_1 \times \cdots \times \gamma_n = \bigcup \tau_i^*(\gamma_i) \in A^*(X^n)$ corresponds to the class $[\underline{\Gamma}] \in A_*(X^n)$.

Now instead of intersecting the cycles $[\nu_i^{-1}(\Gamma_i)]$ in $\overline{M}$ we will consider the product of cohomology classes

$$\underline{\nu}^*(\underline{\gamma}) = \underline{\nu}^*\big(\bigcup \tau_i^*(\gamma_i)\big) = \bigcup \nu_i^*(\gamma_i).$$

We can finally compute the number of points in the intersection in 3.4.3 in terms of such products.

4.1.3 Lemma. *For generic choices of* $\Gamma_1, \ldots, \Gamma_n$ *in* 3.4.3, *the number of points in the intersection* $\underline{\nu}^{-1}(\underline{\Gamma})$ *is equal to*

$$\int [\underline{\nu}^{-1}(\underline{\Gamma})] = \int \underline{\nu}^*(\underline{\gamma}) \cap [\overline{M}].$$

Proof. Recall (cf. Fulton [28], 8.1) that $\underline{\nu}^*(\underline{\gamma}) \cap [\overline{M}]$ is defined as the Gysin pullback $\iota^*([\overline{M} \times \underline{\Gamma}])$, where $\iota : \overline{M} \hookrightarrow \overline{M} \times X^n$ is the graph of $\underline{\nu}$ (which is a regular embedding). Consider the Cartesian diagram

$$\begin{array}{ccc} \underline{\nu}^{-1}(\underline{\Gamma}) & \xrightarrow{\jmath} & \overline{M} \times \underline{\Gamma} \\ {\scriptstyle g}\downarrow & & \downarrow \\ \overline{M} & \xhookrightarrow[\iota]{} & \overline{M} \times X^n. \end{array}$$

The Gysin pullback (cf. [28], 6.1) is now a cycle supported in $\underline{\nu}^{-1}(\underline{\Gamma})$, defined as the intersection of the normal cone $C_\jmath$ with the zero section of the normal bundle g^*N_ι. Since we know that $\underline{\nu}^{-1}(\underline{\Gamma})$ is of correct dimension, it follows that $C_\jmath$ and g^*N_ι have the same dimension. Furthermore, $\underline{\nu}^{-1}(\underline{\Gamma})$ is reduced, and therefore g^*N_ι and consequently $C_\jmath$ are also reduced. It follows that $\iota^*([\overline{M} \times \underline{\Gamma}]) = [\underline{\nu}^{-1}(\underline{\Gamma})]$, as asserted. □

4.1.4 Remark. Assuming that the classes γ_i are Chern classes, we may sketch a simpler proof for the proposition. (This is the case for example when the Γ_i are linear subspaces.) Suppose $\Gamma_i = Z(s_i)$, the zero scheme of a regular section s_i of a vector bundle E_i of rank k_i, so that $\gamma_i = c_{k_i}(E_i)$. Set $\underline{E} := \bigoplus \tau_i^* E_i$ with section $\underline{s} := (s_1, \ldots, s_n)$. Now

$$\bigcap \nu_i^{-1}(\Gamma_i) = \bigcap \nu_i^{-1}(Z(s_i)) = \bigcap Z(\nu_i^* s_i) = Z(\underline{\nu}^* \underline{s}).$$

Knowing that this scheme has correct codimension $k := \sum k_i$, and that $\overline{M}$ is Cohen–Macaulay, we conclude that the section $\underline{\nu}^*\underline{s}$ is regular, and thus its zero scheme is of class $c_k(\underline{\nu}^*\underline{E}) \cap [\overline{M}]$. Now we can write

$$c_k(\underline{\nu}^*\underline{E}) = c_k(\bigoplus \nu_i^* E_i) = \bigcup c_{k_i}(\nu_i^* E_i),$$
$$= \bigcup \nu_i^*\big(c_{k_i}(E_i)\big) = \underline{\nu}^*(\underline{\gamma}),$$

(by naturality) as we wanted.

The above discussion is the enumerative motivation for the following

Definition. The *Gromov–Witten invariant of degree d* associated with the classes $\gamma_1, \ldots, \gamma_n \in A^*(\mathbb{P}^r)$ is

$$I_d(\gamma_1 \cdots \gamma_n) := \int_{\overline{M}} \underline{\nu}^*(\underline{\gamma}).$$

This number is nonzero only when the sum of the codimensions of all the classes γ_i is equal to the dimension of $\overline{M}$.

Note that $I_d(\gamma_1 \cdots \gamma_n)$ is invariant under permutation of the classes γ_i. This is the reason for writing $\gamma_1 \cdots \gamma_n$ with dots as in a product, instead of separating the classes with commas. Note also that since pullback and integration respect sums, the Gromov–Witten invariants are linear in each of their arguments.

The next section and the remainder of this chapter are concerned with the computation of the Gromov–Witten invariants. But first let us record their enumerative interpretation.

4.1.5 Proposition. *Let $\gamma_1, \ldots, \gamma_n \in A^*(\mathbb{P}^r)$ be homogeneous classes of codimension at least 2, with $\sum \operatorname{codim} \gamma_i = \dim \overline{M}_{0,n}(\mathbb{P}^r, d)$. Then for general subvarieties $\Gamma_1, \ldots, \Gamma_n \subset \mathbb{P}^r$ with $[\Gamma_i] = \gamma_i \cap [\mathbb{P}^r]$, the Gromov–Witten invariant $I_d(\gamma_1 \cdots \gamma_n)$ is the number of rational curves of degree d that are incident to all the subvarieties $\Gamma_1, \ldots, \Gamma_n$.*

Proof. The definition combined with Lemma 4.1.3 shows that the Gromov–Witten invariant $I_d(\gamma_1 \cdots \gamma_n)$ is the number of n-pointed stable maps $\mu : \mathbb{P}^1 \to \mathbb{P}^r$ of degree d such that $\mu(p_i) \in \Gamma_i$. In particular, all the rational curves incident to the Γ_i's are in the collection. Now by Lemma 3.5.5 each solution-map μ intersects Γ_i at only one point $\mu(p_i)$. By Lemma 3.5.3 the inverse image of this point is just p_i. Therefore there are no choices left to put the marks. In other words, the number of maps with $\mu(p_i) \in \Gamma_i$ is equal to the number of rational curves incident to the Γ_i's, without mention of marks. □

In particular, the following corollary holds.

4.1.6 Corollary. *For $\mathbb{P}^2$, we have*

$$I_d(\underbrace{h^2 \cdots h^2}_{3d-1\,factors}) = N_d,$$

the number of rational curves of degree d that pass through $3d-1$ general points.

4.1.7 Example. In $\mathbb{P}^3$, the invariant $I_1(h^2 \cdot h^2 \cdot h^2 \cdot h^2)$ is the number of lines incident to four given lines, cf. 3.1.1.

4.1.8 Example. For $\mathbb{P}^3$, the number

$$I_3(\underbrace{h^2 \cdots h^2}_{6} \cdot \underbrace{h^3 \cdots h^3}_{3})$$

is the number of twisted cubics meeting 6 lines and 3 points. It is computed in the space $\overline{M}_{0,9}(\mathbb{P}^3, 3)$. Note that this space has dimension 21, and that this is also the sum of the codimensions of the classes. By the way, the number is 190, as you can compute using the algorithm of Theorem 4.4.1 below.

These Gromov–Witten invariants of twisted cubics were computed already in the 1870s by Schubert [74]. In the proof of 4.4.1, we will come to an algorithm for computing such Gromov–Witten invariants.

4.2 Properties of Gromov–Witten invariants

4.2.1 Lemma. *(Mapping to a point) The only nonzero Gromov–Witten invariants with $d = 0$ are those with 3 marks and $\sum \operatorname{codim} \gamma_i = r$. In this case we have*

$$I_0(\gamma_1 \cdot \gamma_2 \cdot \gamma_3) = \int (\gamma_1 \cup \gamma_2 \cup \gamma_3) \cap [\mathbb{P}^r].$$

Proof. Recall the identification $\overline{M}_{0,n}(\mathbb{P}^r, 0) \simeq \overline{M}_{0,n} \times \mathbb{P}^r$, cf. 2.8.5, and observe that for $n < 3$ this space is empty! Indeed, a constant map $\mathbb{P}^1 \to \mathbb{P}^r$ is unstable unless it has at least three marked points. In the identification, each of the evaluation maps coincides with the projection $\mathrm{pr}_2 : \overline{M}_{0,n} \times \mathbb{P}^r \to \mathbb{P}^r$. Now by definition and by the projection formula we have

$$\begin{aligned} I_0(\gamma_1 \cdots \gamma_n) &= \int_{[\overline{M}]} \nu_1^*(\gamma_1) \cup \cdots \cup \nu_n^*(\gamma_n) \\ &= \int_{[\overline{M}_{0,n} \times \mathbb{P}^r]} \mathrm{pr}_2^*(\gamma_1 \cup \cdots \cup \gamma_n) \\ &= \int \gamma_1 \cup \cdots \cup \gamma_n \cap \mathrm{pr}_{2*}[\overline{M}_{0,n} \times \mathbb{P}^r]. \end{aligned}$$

The projection pr_2 has positive relative dimension and therefore the direct image is zero, unless $n = 3$, so that $\dim \overline{M}_{0,n} = 0$. In this case the last integral above is just $\int \gamma_1 \cup \gamma_2 \cup \gamma_3 \cap [\mathbb{P}^r]$. □

4.2.2 Lemma. *(Two-point invariants) The only nonzero Gromov–Witten invariants with fewer than three marks are*

$$I_1(h^r \cdot h^r) = 1,$$

meaning that there is a unique line passing through two distinct points.

Proof. We can suppose $d > 0$. Then $\dim \overline{M}_{0,n}(\mathbb{P}^r, d) = rd + r + d + n - 3 \geq 2r + n - 2$. Recalling our hypothesis $r \geq 2$, it is clear that for $n < 2$ the sum of the codimensions of the classes γ_i cannot reach $2r + n - 2$. For $n = 2$, the only way is in fact having $d = 1$. □

For the two following lemmas, observe that the diagram

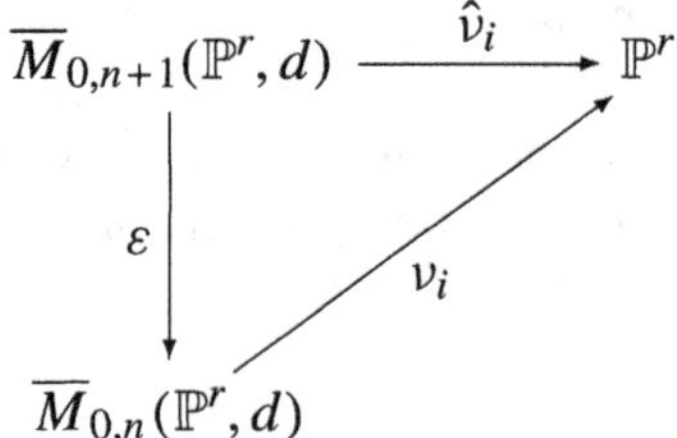

commutes, where ν_i and $\hat{\nu}_i$ are the evaluation maps of the respective spaces; the hat is just to distinguish them. In particular we have the following identity in $A^*\big(\overline{M}_{0,n+1}(\mathbb{P}^r, d)\big)$:

$$\hat{\nu}_i^*(\gamma_i) = \varepsilon^* \nu_i^*(\gamma_i).$$

4.2.3 Lemma. *The only nonzero Gromov–Witten invariants containing a copy of the fundamental class $1 = h^0 \in A^0(\mathbb{P}^r)$ occur in degree zero and with only three marks. (In this case we have $I_0(\gamma_1 \cdot \gamma_2 \cdot 1) = \int \big(\gamma_1 \cup \gamma_2 \cup 1\big) \cap [\mathbb{P}^r]$, as we saw in 4.2.1.)*

Proof. Suppose there is an instance of the fundamental class, say $\gamma_{n+1} = 1$. Note that $\hat{\nu}_{n+1}^*(1) = 1 \in A^*\big(\overline{M}_{0,n+1}(\mathbb{P}^r, d)\big)$. Now whenever $n \geq 3$ or $d > 0$ we can compute the integral by pushdown via ε, using the projection formula:

$$\int \underline{\hat{\nu}}^*(\underline{\gamma}) \cup \hat{\nu}_{n+1}^*(1) \cap [\overline{M}_{0,n+1}(\mathbb{P}^r, d)] \quad = \quad \int \underline{\nu}^*(\underline{\gamma}) \cap \varepsilon_*[\overline{M}_{0,n+1}(\mathbb{P}^r, d)].$$

But $\varepsilon_*[\overline{M}_{0,n+1}(\mathbb{P}^r, d)]$ is zero for dimension reasons. (Note that for $n = 2$ and $d = 0$, there is no forgetful map since the space $\overline{M}_{0,2}(\mathbb{P}^r, 0)$ does not exist!) □

4.2.4 Lemma. *(Divisor equation) Suppose $d > 0$ and that one of the classes is the hyperplane class, say $\gamma_{n+1} = h$. Then*

$$I_d(\gamma_1 \cdots \gamma_n \cdot h) = I_d(\gamma_1 \cdots \gamma_n) \cdot d.$$

Proof. The class $\hat{\nu}^*_{n+1}(h) \cap [\overline{M}_{0,n+1}(\mathbb{P}^r, d)]$ is the class of $\hat{\nu}^{-1}_{n+1}(H)$ for some hyperplane H. It is the locus of maps whose marked point p_{n+1} goes to H. The forgetful map restricted to $\hat{\nu}^{-1}_{n+1}(H)$,

$$\varepsilon_| : \hat{\nu}^{-1}_{n+1}(H) \to \overline{M}_{0,n}(\mathbb{P}^r, d),$$

is generically finite, of degree d. Indeed, for a general map $\mu \in \overline{M}_{0,n}(\mathbb{P}^r, d)$, its image intersects H in d points, and each of the points in the inverse image can acquire the mark p_{n+1}. Now the result follows once again using the projection formula:

$$\begin{aligned}
\int \underline{\hat{\nu}}^*(\underline{\gamma}) \cup \hat{\nu}^*_{n+1}(h) \cap [\overline{M}_{0,n+1}(\mathbb{P}^r, d)] &= \int \underline{\hat{\nu}}^*(\underline{\gamma}) \cap [\hat{\nu}^{-1}_{n+1}(H)] \\
&= \int \underline{\nu}^*(\underline{\gamma}) \cap \varepsilon_*[\hat{\nu}^{-1}_{n+1}(H)] \\
&= \int \underline{\nu}^*(\underline{\gamma}) \cap d\, [\overline{M}_{0,n}(\mathbb{P}^r, d)].
\end{aligned}$$

□

4.2.5 Example. In view of the above properties, when we consider Gromov–Witten invariants, we do not have to worry about those including a factor h^0 (class of $\mathbb{P}^r$) or h^1 (the hyperplane class). In this way, for $\mathbb{P}^2$ it is easy to exhibit them all: the only class to consider is h^2, and to get total codimension equal to $\dim \overline{M}_{0,n}(\mathbb{P}^2, d)$ we need $n = 3d - 1$. In other words, to compute the Gromov–Witten invariants of $\mathbb{P}^2$, it is enough to know

$$I_d(\underbrace{h^2 \cdots h^2}_{3d-1 \text{ factors}}),$$

which are exactly the numbers N_d, cf. Corollary 4.1.6. This is, that the knowledge of all Gromov–Witten invariants of $\mathbb{P}^2$ is equivalent to the information encoded by Kontsevich's formula (together with the lemmas of this section).

4.3 Recursion

Recall that (when $A \neq \emptyset$ and $B \neq \emptyset$) we have the gluing isomorphism 2.7.3.1

$$D(A, B; d_A, d_B) \simeq \overline{M}_{0,A\cup\{x\}}(\mathbb{P}^r, d_A) \times_{\mathbb{P}^r} \overline{M}_{0,B\cup\{x\}}(\mathbb{P}^r, d_B).$$

We are now going to explore this isomorphism to compute integrals over the divisor $D(A, B; d_A, d_B)$ in terms of integrals over the product.

Let us simplify the notation a little. The divisor $D(A, B; d_A, d_B)$ will be denoted by D. We set $\overline{M}_A := \overline{M}_{0,A\cup\{x\}}(\mathbb{P}^r, d_A)$, and write ν_{x_A} for the evaluation map corresponding to the gluing mark $x \in A\cup\{x\}$. Similarly we set $\overline{M}_B := \overline{M}_{0,B\cup\{x\}}(\mathbb{P}^r, d_B)$ with evaluation map ν_{x_B}.

We can express D as the inverse image of the diagonal $\Delta \subset \mathbb{P}^r \times \mathbb{P}^r$:

$$D = (\nu_{x_A} \times \nu_{x_B})^{-1}(\Delta) \subset \overline{M}_A \times \overline{M}_B.$$

Another way to say this is that we have a Cartesian diagram

$$\begin{array}{ccc} D & \overset{\iota}{\hookrightarrow} & \overline{M}_A \times \overline{M}_B \\ \downarrow & & \downarrow \nu_{x_A} \times \nu_{x_B} \\ \mathbb{P}^r & \underset{\delta}{\hookrightarrow} & \mathbb{P}^r \times \mathbb{P}^r \end{array} \qquad (4.3.0.1)$$

where δ is the diagonal embedding. It should be noted that ι is also a regular embedding of the same codimension r.

It is a fundamental fact that we can express the class of the diagonal in terms of the hyperplane classes of the factors:

4.3.1 Künneth decomposition of the diagonal. In the product $\mathbb{P}^r \times \mathbb{P}^r$, with projections π_A and π_B, the class of the diagonal is given by

$$[\Delta] = \sum_{e+f=r} \left(\pi_A^* h^e \cup \pi_B^* h^f\right) = \sum_{e+f=r} (h^e \times h^f).$$

These sums over $e + f = r$ will appear throughout; it is always understood of course that e and f are nonnegative integers.

The Künneth formula is a consequence of the well-known fact that the diagonal is the zero scheme of a regular section of the vector bundle

$$E = \pi_A^* T_{\mathbb{P}^r}(-1) \otimes \pi_B^* \mathcal{O}(1),$$

which in turn follows by pulling back the Euler sequence. (See [2], Prop. (2.7), p. 18.) Now the right-hand side of the expression of $[\Delta]$ above is just the expansion of the rth Chern class of the rank-r vector bundle E.

Since $\overline{M}_A \times \overline{M}_B$ is Cohen–Macaulay and D is of the correct codimension, this allows us to write the class of D in the following way.

$$\begin{aligned} [D] &= (\nu_{x_A} \times \nu_{x_B})^*[\Delta] \\ &= (\nu_{x_A} \times \nu_{x_B})^*\Big(\sum_{e+f=r} (h^e \times h^f) \cap (X \times X)\Big) \\ &= \sum_{e+f=r} \left(\nu_{x_A}^* h^e \times \nu_{x_B}^* h^f\right) \cap (\overline{M}_A \times \overline{M}_B). \end{aligned}$$

Finally, we can state the key lemma, often called the Splitting Lemma.

4.3.2 Lemma. *(Splitting Lemma) Let* $\alpha : D \hookrightarrow \overline{M}$ *be the natural inclusion, and let* $\iota : D \hookrightarrow \overline{M}_A \times \overline{M}_B$ *be the inclusion described above. Then for any classes* $\gamma_1, \ldots, \gamma_n \in A^*(\mathbb{P}^r)$ *the following identity holds in* $A^*(\overline{M}_A \times \overline{M}_B)$*:*

$$\iota_* \alpha^* \underline{\nu}^*(\underline{\gamma}) = \sum_{e+f=r} \Big(\prod_{a \in A} \nu_a^*(\gamma_a) \cdot \nu_{x_A}^*(h^e)\Big) \times \Big(\prod_{b \in B} \nu_b^*(\gamma_b) \cdot \nu_{x_B}^*(h^f)\Big).$$

Proof. The key point is simply the compatibility between evaluation maps and the recursive structure, cf. 2.8.1: thus the restriction to D of an evaluation class $\nu_i^*\gamma_i$ gives the evaluation class of the same mark p_i on the moduli space corresponding to the twig containing p_i. So in the situation of the lemma, all the classes corresponding to the marks in A become classes on $\overline{M}_A$ and all the classes corresponding to B become classes on $\overline{M}_B$. But then there are some new cohomology classes at the gluing marks which express the fact that D is not the whole product $\overline{M}_A \times \overline{M}_B$ but only a subvariety in there, given as inverse image of the diagonal, wherein we use Künneth.

First some notation, similar to what we have already used: $X := \mathbb{P}^r$ and $\underline{X} = X \times \cdots \times X$ (n copies). Denote by $\underline{X}_A$ the partial product of the factors indexed by A, and similarly for $\underline{X}_B$. Hence, $\underline{X} = \underline{X}_A \times \underline{X}_B$. Let $\underline{\nu} : \overline{M} \to \underline{X}$ be the product of the n evaluation maps $\overline{M} \to X$. Let $\underline{\nu}_A : \overline{M}_A \to \underline{X}_A$ be the product of the evaluation maps corresponding to the marks in A and define similarly $\underline{\nu}_B : \overline{M}_B \to \underline{X}_B$. Note that we do not include the evaluation map of the gluing mark x. Finally, let $\underline{\gamma}$ be the class $\gamma_1 \times \cdots \times \gamma_n$ in $A^*(\underline{X})$, and let $\underline{\gamma}_A \in A^*(\underline{X}_A)$ and $\underline{\gamma}_B \in A^*(\underline{X}_B)$ be defined in the obvious way. (The philosophy of notation should be clear by now.) Note that $\underline{\nu}^*(\underline{\gamma}) = \nu_1^*(\gamma_1) \cup \cdots \cup \nu_n^*(\gamma_n)$.

Having agreed on these notations we can write the following commutative diagram, which just expresses the compatibility:

$$\begin{array}{ccc} \overline{M} & \xrightarrow{\ \underline{\nu}\ } & \underline{X} \\ \uparrow{\scriptstyle \alpha} & & \uparrow{\scriptstyle \underline{\nu}_A \times \underline{\nu}_B} \\ D & \xrightarrow[\ \iota\ ]{} & \overline{M}_A \times \overline{M}_B \end{array}$$

Thus $\alpha^* \underline{\nu}^*(\underline{\gamma}) = \iota^*(\underline{\nu}_A \times \underline{\nu}_B)^*(\underline{\gamma})$.

We now push this class into $\overline{M}_A \times \overline{M}_B$ along ι. In the case at hand this means that for $z \in A_*(\overline{M}_A \times \overline{M}_B)$ we may write

$$\begin{aligned} \iota_*\big(\iota^*(\underline{\nu}_A \times \underline{\nu}_B)^*(\underline{\gamma})\big) \cap z &= \iota_*\big(\iota^*(\underline{\nu}_A \times \underline{\nu}_B)^*(\underline{\gamma}) \cap \iota^* z\big) \\ &= \iota_*\Big(\iota^*\Big((\underline{\nu}_A \times \underline{\nu}_B)^*(\underline{\gamma}) \cap z\Big)\Big) \\ &= (\underline{\nu}_A \times \underline{\nu}_B)^*(\underline{\gamma}) \cap c_r((\nu_{x_A} \times \nu_{x_B})^* E) \cap z, \end{aligned}$$

and from here, just expand the Chern classes of E as before by the Künneth decomposition as $\sum_{e+f=r} \nu_{x_A}^* h^e \times \nu_{x_B}^* h^f$.

So altogether,

$$\iota_* \alpha^* \underline{\nu}^*(\underline{\gamma}) = (\underline{\nu}_A \times \underline{\nu}_B)^*(\underline{\gamma}) \cup \Big(\sum_{e+f=r} \nu_{x_A}^* h^e \times \nu_{x_B}^* h^f \Big).$$

The conclusion follows by separating the classes according to which moduli space they are pulled back from. □

Integrating, we obtain the following.

4.3.3 Corollary.

$$\int_D \nu_1^*(\gamma_1) \cup \cdots \cup \nu_n^*(\gamma_n) = \sum_{e+f=r} I_{d_A}\Big(\prod_{a\in A} \gamma_a \cdot h^e\Big) \cdot I_{d_B}\Big(\prod_{b\in B} \gamma_b \cdot h^f\Big).$$

4.4 The reconstruction theorem

We now show that it is possible to reduce the computation of any I_d to the single one $I_1(h^r \cdot h^r) = 1$.

As a warm-up, let us see how it works in the example 3.2.2 already treated: in $\mathbb{P}^2$, we shall retrieve $N_2 = I_2(2,2,2,2,2) = 1$, the number of conics passing through 5 general points. Once again we place ourselves in the space $\overline{M}_{0,6}(\mathbb{P}^2, 2)$. Take six classes $\lambda_1 = \lambda_2 = h$ and $\gamma_1 = \gamma_2 = \gamma_3 = \gamma_4 = h^2$, and let the six marks be denoted by $m_1, m_2, p_1, \ldots, p_4$. In analogy with 3.2.2, consider the class

$$\underline{\nu}^*(\underline{\gamma}) = \nu_{m_1}^*(\lambda_1) \cup \nu_{m_2}^*(\lambda_2) \cup \nu_{p_1}^*(\gamma_1) \cup \nu_{p_2}^*(\gamma_2) \cup \nu_{p_3}^*(\gamma_3) \cup \nu_{p_4}^*(\gamma_4)$$

in $A^{10}\big(\overline{M}_{0,6}(\mathbb{P}^2, 2)\big)$. (Taking $\underline{\nu}^*(\underline{\gamma}) \cap [\overline{M}]$, we obtain exactly the class of the curve Y constructed on page 95.) Next, intersect this curve with the two equivalent special boundary divisors, getting

$$\int \underline{\nu}^*(\underline{\gamma}) \cap D(m_1, m_2 | p_1, p_2) = \int \underline{\nu}^*(\underline{\gamma}) \cap D(m_1, p_1 | m_2, p_2).$$

As we also did in the proof of Proposition 3.2.2, we compute the contribution from each component of the divisors. The left-hand side is

$$\sum \Big(\int \underline{\nu}^*(\underline{\gamma}) \cap D(A, B; d_A, d_B) \Big)$$

where the sum is over all partitions $A \cup B = \{m_1, m_2, p_1, p_2, p_3, p_4\}$, with $m_1, m_2 \in A$ and $p_1, p_2 \in B$ and with weights $d_A + d_B = 2$.

At this point enters the Splitting Lemma, and more precisely its Corollary 4.3.3. It allows us to write the last integral as

$$\sum\left(\sum_{e+f=2} I_{d_A}(\lambda_1\cdot\lambda_2\cdot\prod\gamma_a\cdot h^e)I_{d_B}(\gamma_1\cdot\gamma_2\cdot\prod\gamma_b\cdot h^f)\right)$$

where the outer sum is over the same data as above, and the products come from the various manners of distributing the two spare marks in A and B. The Gromov–Witten invariants are computed over moduli spaces with marking sets $A \cup \{x\}$ and $B \cup \{x\}$ respectively.

Once again, we analyze which choices of weights might give any contribution. Suppose $d_A = 0$. Then by the above observation, there can be only three marks in $A \cup \{x\}$, to wit, m_1, m_2 and x. For the three corresponding classes to be of correct codimension we must have $e = 0$. Then $I_0(\lambda_1 \cdot \lambda_2 \cdot h^0) = 1$, and the second factor is $I_2(\gamma_1 \cdot \gamma_2 \cdot \gamma_3 \cdot \gamma_4 \cdot h^2) = N_2$, exactly the sought-for invariant. There is no contribution from $d_B = 0$ because in the Gromov–Witten invariants indexed by B the codimension is already too big due to the presence of γ_1 and γ_2.

Finally, we must examine the contributions with $d_A = d_B = 1$. They are given by

$$\sum\left(\sum_{e+f=2} I_1(\lambda_1\cdot\lambda_2\cdots h^e)I_1(\gamma_1\cdot\gamma_2\cdots h^f)\right),$$

where $\cdots$ represents the various possible ways to distribute the spare marks. Let us check the possibilities one by one. If there are no spare marks in A, then the number of marks in $A \cup \{x\}$ is 3, and $\dim \overline{M}_{0,A\cup\{x\}}(\mathbb{P}^2, 1) = 5$. On the other hand, the total codimension of the corresponding classes is $2 + e$. So no contribution occurs in this way. Suppose there is one spare mark in A and another in B. Once again we compare the dimension of the space in question with the total codimension of the classes. We see that the Gromov–Witten invariants at play all vanish. Finally, putting both spare marks on the A-twig, for $e = f = 1$ we find the contribution given by

$$I_1(\lambda_1\cdot\lambda_2\cdot\gamma_3\cdot\gamma_4\cdot h^1)I_1(\gamma_1\cdot\gamma_2\cdot h^1).$$

In order to compute these Gromov–Witten invariants, recall (from 4.2.4) that the pure h's can be thrown outside (in exchange for a degree, but here $d = 1$). Thus the above expression becomes equal to

$$I_1(\gamma_3\cdot\gamma_4)I_1(\gamma_1\cdot\gamma_2) = 1\cdot 1.$$

In either case this number 1 is interpreted as the number of lines passing through two distinct points.

Similarly we can find an expression for the right-hand side. Here the important thing to notice is that neither $d_A = 0$ nor $d_B = 0$ gives a contribution. This is easy to see, because in either case on the twig in question there can be only three marked points and then the codimensions would become too big, already due to the classes λ and γ.

4.4.1 Theorem. (Reconstruction for $\mathbb{P}^r$) (Kontsevich–Manin [58] and Ruan–Tian [73]) *All the (genus-0) Gromov–Witten invariants for $\mathbb{P}^r$ can be computed recursively, and the only necessary initial value is $I_1(h^r \cdot h^r) = 1$, the number of lines through two points.*

Proof. (*Sketch*) The recursion for $\mathbb{P}^r$ is not as direct as the one we saw in the case of $\mathbb{P}^2$. It is given by a huge collection of highly redundant equations. Let us outline the algorithm. Recall that the only invariant with 2 marks is $I_1(h^r \cdot h^r) = 1$. Therefore, to prove that the recursion terminates we need to express each Gromov–Witten invariant in terms of invariants of lower degree, or with the same degree but with fewer marks.

If a class of codimension zero occurs, use Lemma 4.2.3 to get rid of it. If a class of codimension 1 occurs, use Lemma 4.2.4 to dispose of it in exchange for a degree. Hence we can suppose that all the classes are of codimension at least 2. Let us rearrange the classes so that the ones of highest codimension come first and the ones of lowest codimension last. Write the last class as $\gamma_n = \lambda_1 \cup \lambda_2$, where each of these new classes is of codimension strictly smaller than the codimension of γ_n. (This is possible since h generates the cohomology ring $A^*(\mathbb{P}^r)$.)

Now the computation is performed in the space $\overline{M}_{0,n+1}(\mathbb{P}^r, d)$. Let us denote the marks by $m_1, m_2, p_1, \ldots, p_{n-1}$. Consider the class

$$\nu_{m_1}^*(\lambda_1) \cup \nu_{m_2}^*(\lambda_2) \cup \nu_{p_1}^*(\gamma_1) \cup \ldots \cup \nu_{p_{n-1}}^*(\gamma_{n-1}),$$

which is the class of a curve. (Note that this is exactly what we did for conics.) Integrate this class over the two equivalent special boundary divisors, $D(m_1, m_2|p_1, p_2)$ and $D(m_1, p_1|m_2, p_2)$, in other words, intersect the curve with each of these divisors. Applying the Splitting Lemma yields an equation involving several Gromov–Witten invariants, all of type

$$I_{d_A}\Big(\prod_{a\in A} \gamma_a \cdot h^e\Big) I_{d_B}\Big(\prod_{b\in B} \gamma_b \cdot h^f\Big).$$

Here the products are taken over all the classes indexed by marks belonging to A or B, respectively. The classes h^e and h^f correspond to the gluing mark x. Now if both d_A and d_B are strictly positive, the Gromov–Witten invariants are known by the induction hypothesis, since they have lower degree.

To see that the algorithm terminates, we have to examine the contribution from $d_A = 0$ and $d_B = 0$. We know (cf. 4.2.1) that the Gromov–Witten invariants of degree 0 are those with three marks. Therefore, only the following four terms survive:

$$I_0(\lambda_1 \cdot \lambda_2 \cdot h^{r-c_1-c_2}) I_d(\gamma_1 \cdot \gamma_2 \cdot \gamma_3 \cdots \gamma_{n-1} \cdot h^{c_1+c_2}),$$
$$I_0(\lambda_1 \cdot \gamma_1 \cdot h^{r-c_1-b_1}) I_d(h^{b_1+c_1} \cdot \gamma_2 \cdot \gamma_3 \cdots \gamma_{n-1} \cdot \lambda_2),$$
$$I_d(\gamma_1 \cdot h^{c_2+b_2} \cdot \gamma_3 \cdots \gamma_{n-1} \cdot \lambda_1) I_0(\lambda_2 \cdot \gamma_2 \cdot h^{r-c_2-b_2}),$$
$$I_d(h^{b_1+b_2} \cdot \lambda_2 \cdot \gamma_3 \cdots \gamma_{n-1} \cdot \lambda_1) I_0(\gamma_1 \cdot \gamma_2 \cdot h^{r-b_1-b_2}),$$

where $c_i = \operatorname{codim} \lambda_i$ and $b_i = \operatorname{codim} \gamma_i$.

The I_0-factors are all equal to 1, so the first of the four terms is exactly the invariant $I_d(\gamma_1 \cdots \gamma_n)$ we were looking for. The other three terms have a λ_i-factor of lower codimension at the end. So we can pass them on to recursion: eventually the last factor achieves codimension 1, and then we can dispose of it as in 4.2.4. Thus we have a term with fewer marks. Continuing like this we eventually arrive at the situation where there are only two marks, but the only nonzero such invariant is $I_1(h^r \cdot h^r) = 1$. □

A MAPLE implementation of the algorithm just described is available from the home page of this book. A fancier program farsta was written by Kresch [59].

4.5 Generalizations and references

4.5.1 Gromov–Witten invariants of convex varieties are defined as for $\mathbb{P}^r$, and behave similarly. The computation for $\mathbb{P}^3$ and for the smooth quadric three-fold is performed in FP-NOTES. In the pioneering paper of Di Francesco and Itzykson [17] there are other examples like $\mathbb{P}^1 \times \mathbb{P}^1$ and $\mathrm{Gr}(1, \mathbb{P}^3)$ (although the notion of Gromov–Witten invariant is not explicit). Another interesting example is provided by Ernström and Kennedy [23]. They define a space of *stable lifts* for $\mathbb{P}^2$ which is a subspace of the space of stable maps to the incidence variety I of points and lines in $\mathbb{P}^2$. This space codifies second order information like tangency and cusp behavior.

For such varieties, an additional complication is that the initial value for the recursion is no longer as simple as the case "a unique line through two distinct points" of $\mathbb{P}^r$ (see 4.5.4). For example, in the case of the incidence variety of Ernström and Kennedy [23], six initial values are needed.

4.5.2 Virtual fundamental class. For nonconvex varieties and for higher-genus stable maps we noticed in 2.10.2 and 2.10.3 that the moduli space has components of excessive dimension, and it is not at all obvious that a reasonable kind of intersection theory can work to define the Gromov–Witten invariants. Amazingly, such a theory exists and the Gromov–Witten invariants can be defined for any smooth projective variety in a way that make the basic properties work, to wit, mapping to a point 4.2.1, string equation 4.2.3, divisor equation 4.2.4, and compatibility with the structure maps, including the Splitting Lemma 4.3.2.

The crucial notion is that of the *virtual fundamental class*, to use in place of $[\overline{M}_{g,n}(X, \beta)]$. If the expected dimension of $\overline{M}_{g,n}(X, \beta)$ is s, then the virtual fundamental class lives in $A_s(\overline{M}_{g,n}(X, \beta))$. The relation between the virtual class and the space itself is a lot like the relation between the top Chern class of a vector bundle and the zero scheme of a section of it: even if the section is not regular and defines a scheme with components of excessive dimension, the top Chern class lives in correct dimension (see Fulton [28, 14.1]). The construction of this class is very technical and depends on a lot of deformation theory, and it is more naturally expressed in the language of stacks. The interested reader is referred to the original paper of Behrend and Fantechi [7].

In the general theory of stable maps and Gromov–Witten invariants, this virtual fundamental class with its nice properties is really the central notion. In our present setup, $X = \mathbb{P}^r$, all this theory is hidden due to the fact that the virtual class simply coincides with the usual topological fundamental class in this case.

4.5.3 Nonconvex varieties. Both the definition and the computation of Gromov–Witten invariants of a nonconvex variety X require the use of the virtual fundamental class. In this case, the Gromov–Witten invariants do not afford direct enumerative interpretation in general. But sometimes the Gromov–Witten invariants of X can be interpreted as counting curves on related varieties. For example, in many cases the Gromov–Witten invariants of projective spaces (notably $\mathbb{P}^2$) blown up at points can be interpreted as the numbers of rational curves in $\mathbb{P}^r$ with prescribed multiple points at the blowup centers, cf. Göttsche and Pandharipande [36] and Gathmann [31]. Another example is when X is the Hilbert scheme of 2 points in $\mathbb{P}^2$. Here certain (genus-zero) Gromov–Witten invariants can be interpreted as the number of hyperelliptic curves (higher genus) passing through the appropriate number of points; cf. Graber [37].

4.5.4 Reconstruction in general. A key step in the reconstruction argument was decomposing the last class as $\gamma_n = \lambda_1 \cup \lambda_2$. Eventually one of these classes would be a divisor, which we could then take away using 4.2.4. In general for this argument to work we need to assume that $H^*(X)$ is generated by divisor classes (over $\mathbb{Q}$). For $\mathbb{P}^r$, the only needed initial input was $I_1(h^r \cdot h^r) = 1$. In general reconstruction will determine all the Gromov–Witten invariants from those with $n \leq 2$.

But in cases like for instance the quintic three-fold in $\mathbb{P}^4$, where the virtual dimension is 0 in every degree (cf. 2.10.2) the only invariants are the 0-pointed invariants $I_d()$ (modulo the fact that you can always throw in divisor classes, cf. 4.2.4). So in this case, reconstruction does not provide any information at all.

4.5.5 Gravitational descendants. An important generalization Gromov–Witten invariants is the notion of *gravitational descendants*, or descendant Gromov–Witten invariants. While the Gromov–Witten invariants depend solely on pullbacks of classes of $\mathbb{P}^r$, the gravitational descendants involve also the psi classes which can be defined for stable maps just as for stable curve (cf. 1.6.6): $\psi_i := c_1(\sigma_i^* \omega_\pi)$, where ω_π is the relative dualizing sheaf of the forgetful map $\pi : \overline{M}_{0,n+1}(\mathbb{P}^r, d) \to \overline{M}_{0,n}(\mathbb{P}^r, d)$ and σ_i is the section corresponding to mark p_i (cf. 1.5.11). The fiber of $\sigma_i^* \omega_\pi$ at a moduli point $[\mu : C \to \mathbb{P}^r]$ is the cotangent space $(T_{p_i} C)^*$.

The *gravitational descendants* are by definition products of psi classes and classes pulled back from $\mathbb{P}^r$. The name comes from physics, where including the psi classes corresponds to coupling the field theory to gravity; cf. [86]. Psi classes are central to most deep results in Gromov–Witten theory, including all applications to mirror symmetry (see Pandharipande [67]).

The psi classes also play an important role in treating tangency conditions, higher contacts, as well as other types of infinitesimal behavior (see Graber–Kock–Pandharipande [38], Gathmann [32], Kock [53]).

4.5.6 Tree-level systems and CohFT structures. Let us just mention the structure of a *cohomological field theory* (CohFT) which is a generalization of the notion of Gromov–Witten invariants.

Instead of looking only at top intersections as in the case of Gromov–Witten invariants, one can look at more general cohomology classes. In other words, start out with any collection of cohomology classes of X; take their pullbacks to $\overline{M}_{0,n}(X, \beta)$ via evaluation maps; and now instead of integrating, take the direct image in $\overline{M}_{0,n}$ via the forgetful map $\eta : \overline{M}_{0,n}(X, \beta) \to \overline{M}_{0,n}$ (cf. 2.6.6). This gives, for each $n \geq 3$, a map

$$\begin{aligned} I^X_{n,d} : A^*(X)^{\otimes n} &\longrightarrow A^*(\overline{M}_{0,n}) \\ \underline{\gamma} &\longmapsto \eta_*\big(\underline{\nu}^*(\underline{\gamma})\big). \end{aligned}$$

This collection of maps is called a *tree-level system*; cf. Kontsevich–Manin [58] (because it involves only genus zero where all curves are trees). In the same way as integration over equivalent boundary divisors in $\overline{M}_{0,n}(X, \beta)$ yields the Splitting Lemma 4.3.2 in the case of Gromov–Witten invariants, intersection with equivalent divisors of type $D = D(A|B) \subset \overline{M}_{0,n}$ yields a similar recursive relation which compares $I^X_{n,d}$ with the $I^X_{n_A+1,d_A}$, $I^X_{n_B+1,d_B}$ of the twigs.

A second reconstruction theorem of Kontsevich–Manin [58] says that the Gromov–Witten invariants determine the whole tree-level system. Roughly, this is a consequence of the result that $A^*(\overline{M}_{0,n})$ is spanned by boundary classes.

More generally, any collection of multilinear maps $A^*(X)^{\otimes n} \to A^*(\overline{M}_{0,n})$ invariant under permutation and obeying a recursion like the one indicated above is called a *cohomological field theory* (CohFT). See the book of Manin [61]. It does not have to be defined just with pullback classes; other classes can be included as well, for example psi classes.

4.5.7 Higher genus. It should be mentioned that Gromov–Witten invariants are defined in higher genus just as in genus zero, but there is no reconstruction algorithm like the one of 4.4.1; there is no linear equivalence like 2.7.5.1, which is the crucial point in the reconstruction described.

Some special cases of genus-1 recursion exist [34]. Otherwise the known relations (including the famous Virasoro constraints, which are known to hold in some cases [35]) are all in the setting of gravitational descendants. The most successful methods for computing Gromov–Witten invariants in higher genus rely on torus actions and variations on Bott's localization formula. The starting point for these developments is Kontsevich [57]; the standard reference is Graber–Pandharipande [39].

Exercises

3-point invariants

1. Generalizing Lemma 4.2.2, give a characterization of all possible nonzero Gromov–Witten invariants of $\mathbb{P}^r$ with 3 marks. (You are going to compute them in Exercise 4.)

2. Use the reconstruction algorithm to compute the Gromov–Witten invariant $I_1(h^3, h^2, h^2) = 1$ for $\mathbb{P}^3$, which is the number of lines through one point and two lines.

3. Use the reconstruction algorithm to compute the following two Gromov–Witten invariants for $\mathbb{P}^4$: the number of lines through a point, a line, and a plane: $I_1(h^4, h^3, h^2) = 1$, and the number of lines through 3 given lines, $I_1(h^3, h^3, h^3) = 1$.

4. Generalizing the two previous exercises, show that all nonzero 3-point invariants of $\mathbb{P}^r$ are equal to 1.

Gromov–Witten invariants of $\mathbb{P}^1$

5. Lemma 4.2.2 assumes $r \geq 2$. Show that the conclusion is false for $\mathbb{P}^1$; state and prove the correct result in this case.

6. Use dimension constraints and the divisor equation 4.2.4 to compute all Gromov–Witten invariants of $\mathbb{P}^1$. In positive degree they are

$$I_1(h^{\bullet n}) = 1, \qquad n \geq 0.$$

7. Let D be any boundary divisor in $\overline{M}_{0,n}(\mathbb{P}^1, 2)$. Use Corollary 4.3.3 and the previous exercise to show that

$$\int_D \nu_1^* \gamma_1 \cup \cdots \cup \nu_n^* \gamma_n = 0$$

for any n-tuple of cohomology classes γ_i.

The Grassmannian $G = G(2,4)$. We use the standard basis for the cohomology of the Grassmannian $G = G(2,4)$ of lines in $\mathbb{P}^3$:

	dim	lines in $\mathbb{P}^3$
T_0	4	all lines
T_1	3	lines meeting a line
T_2	2	lines through a point
T_3	2	lines in a plane
T_4	1	lines in a plane through a point
T_5	0	fixed line

By a degree-d curve in G we mean a curve of class d times T_4.

8. Show that the only nonzero 2-point invariant for G is

$$I_1(T_4 T_5) = 1.$$

Compute this invariant by a direct geometric argument in $\mathbb{P}^3$: a degree-1 curve C in G is just a linear pencil of lines in $\mathbb{P}^3$, i.e., the set of lines contained in a given plane Π and passing through a given point $p \in \Pi$. There are two conditions imposed: C should meet a given degree-1 curve L (i.e., a cycle of degree T_4) and pass through a point $q \in G$ (i.e., a cycle of degree T_5). The first condition means the pencil needs to have a line in common with a given pencil (corresponding to L), and the second condition means it must contain a given line (corresponding to q). Construct the unique solution from these conditions.

9. Show that by dimension constraints, the only possible nonzero three-point invariants for $G(2,4)$ are

$$I_1(T_2T_2T_5), \quad I_1(T_2T_3T_5), \quad I_1(T_3T_3T_5), \quad I_1(T_1T_4T_5), \qquad I_2(T_5T_5T_5).$$

Hint: the dimension of the moduli space was computed in Exercise 17 on page 90.

10. (i) Show by elementary geometric arguments in $\mathbb{P}^3$ that we have

$$I_1(T_2T_3T_5) = 1$$

$$I_1(T_2T_2T_5) = I_1(T_3T_3T_5) = 0.$$

Hint for the first one (cf. the interpretation used in Exercise 8 on the page before): we are given a line a plane and a point in $\mathbb{P}^3$, and we need to find a pencil of lines that contains the given line, contains a line through the given point, and contains a line in the given plane. The arguments for the next two are similar.

(ii) Show also that $I_1(T_1T_4T_5) = 1$. (You almost did it already in Exercise 8 on the preceding page.)

11. A rational curve C on G of degree d parametrizes a family of lines in $\mathbb{P}^3$: these sweep out a rational ruled surface S of degree d.

(i) Show that the condition on C of being incident to a cycle of class T_2 (lines through a given point) corresponds to the condition on S of containing that point, and that the condition on C of passing through a point $q \in G$ corresponds to the condition on S of containing the line in $\mathbb{P}^3$ corresponding to q.

(ii) Show that $I_2(T_2^{\bullet 9}) = 2$. (*Hint:* the quadrics in $\mathbb{P}^3$ form a $\mathbb{P}^9$, and each point is a linear condition. But each quadric is a $\mathbb{P}^1$-bundle on $\mathbb{P}^1$ in two ways.)

(iii) Show that $I_2(T_5T_5T_5) = 1$, by interpreting the number as a count of quadrics containing three lines.

(iv) Show that the condition on C of being incident to a cycle of class T_3 (lines in a given plane) corresponds to the condition on S of being tangent to the corresponding plane.

(v) Argue that the numbers $I_2(T_2^{\bullet a}T_3^{\bullet b})$ are the characteristic numbers of quadric surfaces in $\mathbb{P}^3$, i.e., the numbers of quadric surfaces incident to a points and tangent to b planes, $a + b = 9$.

Chapter 5

Quantum Cohomology

In this final chapter we will construct the Gromov–Witten potential, which is the generating function for the Gromov–Witten invariants, and use it to define a *quantum product* on $A^*(\mathbb{P}^r)$. Kontsevich's formula and the other recursions we found in Chapter 4, are then interpreted as partial differential equations for the Gromov–Witten potential. The striking fact about all these equations is that they amount to the associativity of the quantum product! In particular, Kontsevich's formula is equivalent to associativity of the quantum product of $\mathbb{P}^2$.

Since the formalism of generating functions is not an everyday tool for most algebraic geometers, we start with a very short introduction to this subject; hopefully this will render the manipulations with the Gromov–Witten potential less magic.

5.1 Quick primer on generating functions

The technique of generating functions is a very useful tool for managing sequences or arrays of numbers, especially if the numbers are related by recursive relations. One standard reference is Stanley's *Enumerative Combinatorics* [75]; another option is Wilf's *generatingfunctionology* [85] (which is freely available on the Internet!).

5.1.1 Generating functions. Suppose we are given a sequence of numbers $\{N_k\}_{k=0}^{\infty}$; typically the numbers N_k count something that depends on k. The idea is to store all the numbers as coefficients in a formal power series called the generating function

$$F(x) := \sum_{k=0}^{\infty} \frac{x^k}{k!} N_k.$$

We are not so interested in the function aspect of this object: we do not care what its value is for $x = 7$, or whether it is convergent. It is merely a data structure for holding all the numbers N_k. The formal variable x does the job of distinguishing the terms so that we can extract the numbers from F, since (by definition) two formal power series are equal if and only if their coefficients of x^k are equal for all k. The factorials $1/k!$ are sometimes omitted in the definition, but they are convenient for our purposes (F is more precisely called an exponential generating function).

The point is that properties of the sequence $\{N_k\}_{k=0}^{\infty}$ often can be expressed in terms of properties of F, and in particular, recursive relations among the numbers N_k translate into differential equations for F. To see this, two important observations are in place.

5.1.2 Derivatives of a generating function. First, *the formal derivative* $F_x := \frac{d}{dx}F$ *is the generating function for the sequence* $\{N_{k+1}\}_{k=0}^{\infty}$, *in the sense that*

$$F_x = \frac{d}{dx}F = \sum_{k=0}^{\infty} \frac{x^k}{k!} N_{k+1}.$$

The proof is a trivial exercise: $\frac{d}{dx}F = \sum_{k=1}^{\infty} k\frac{x^{k-1}}{k!} N_k = \sum_{k=1}^{\infty} \frac{x^{k-1}}{(k-1)!} N_k = \sum_{k=0}^{\infty} \frac{x^k}{k!} N_{k+1}$. In the last step we simply shifted the running index.

To see how it works, let us briefly consider a simple example.

5.1.3 Example. *(Fibonacci numbers)* Recall that the Fibonacci numbers $\{N_k\}_{k=0}^{\infty}$ are defined recursively by the initial condition $N_0 = N_1 = 1$ together with the relation

$$N_{k+2} = N_{k+1} + N_k, \qquad k \geq 0.$$

Let F denote the generating function, $F = \sum_{k=0}^{\infty} \frac{x^k}{k!} N_k$. The collection of all these recursive relations amounts to an equation of sequences $\{N_{k+2}\}_{k=0}^{\infty} = \{N_{k+1} + N_k\}_{k=0}^{\infty}$, which in turn (by the above observation) is equivalent to the following differential equation for F:

$$F_{xx} = F_x + F.$$

The conditions $N_0 = N_1 = 1$ translate into initial conditions for the differential equation: $F(0) = 1$ and $F_x(0) = 1$.

(In fact one can actually solve this differential equation and obtain $F(x) = \frac{e^{r_1 x} - e^{r_2 x}}{\sqrt{5}}$, where $r_i = \frac{1 \pm \sqrt{5}}{2}$, but this is not our point here.)

5.1.4 Product rule for generating functions. Second observation: Suppose we are given two generating functions $F(x) = \sum_{k=0}^{\infty} \frac{x^k}{k!} f_k$ and $G(x) = \sum_{k=0}^{\infty} \frac{x^k}{k!} g_k$, for some numbers f_k and g_k. Then *the product* $F \cdot G$ *is the generating function for the numbers* $h_k := \sum_{i=0}^{k} \binom{k}{i} f_i \, g_{k-i}$.

The proof is just a matter of multiplying power series:

$$\begin{aligned} F \cdot G &= \Big(\sum_{i=0}^{\infty} \frac{x^i}{i!} f_i\Big)\Big(\sum_{j=0}^{\infty} \frac{x^j}{j!} g_j\Big) \\ &= \sum_{i,j} \frac{x^{i+j}}{i!\,j!} f_i\, g_j \\ &= \sum_{k=0}^{\infty} x^k \Big(\sum_{i+j=k} \frac{f_i\, g_j}{i!\,j!}\Big) \\ &= \sum_{k=0}^{\infty} \frac{x^k}{k!} \Big(\sum_{i=0}^{k} \tbinom{k}{i} f_i\, g_{k-i}\Big). \end{aligned}$$

Intuitively, what is going on is that the counting problem is cut into two pieces. The sum over i and the binomial factor $\binom{k}{i}$ represent all the ways we can cut the problem in two, and then in one part we are left with the count f_i and in the other part we have g_{k-i}.

5.1.5 Example. *(Bell numbers)* The kth Bell number N_k is the number of ways to partition a k-element set S into disjoint nonempty subsets (and we define $N_0 = 1$). They satisfy the recurrence

$$N_{k+1} = \sum_{i=0}^{k} \binom{k}{i} N_{k-i},$$

as easily follows from this argument: classify the partitions of $S \cup \{p_0\}$ according to the number i of elements that are in the same part as p_0. There are $\binom{k}{i}$ choices for which elements go with p_0. For each of these choices there is only one way to group these elements together with p_0 (so there is an invisible factor 1 in the formula), while there are N_{k-i} ways to partition the remaining elements.

The recurrence translates into an equation for the generating function $F(x) = \sum_{k=0}^{\infty} \frac{x^k}{k!} N_k$. The left side of the equation corresponds to the generating function F_x, according to 5.1.2. On the right-hand side, the omitted factor 1 can be considered as the entries of the constant sequence (whose generating function is clearly $e^x = \sum_{k=0}^{\infty} \frac{x^k}{k!}$), so we recognize the number on the right-hand side as the coefficients in the generating function $e^x F$, according to the product rule. Thus we get the differential equation

$$F_x = e^x F$$

(which incidentally can be solved (using the initial condition, $F(0) = 1$): $F = \exp(e^x - 1)$.)

There are many things one can do with these generating functions; cf. *op. cit.* What we did was nearly nothing. We just translated a recursion among numbers into a differential equation for their generating function, and then we were lucky: In each case we could make an interesting remark about the differential equation, saying, "Aha! this is the differential equation satisfied by ... ." We are going to do the same for the Gromov–Witten invariants: we will organize the recursions of Chapter 4 into a differential equation for the generating function for the Gromov–Witten invariants, and then we are going to say, "Aha! this differential equation is precisely the expression for associativity of a certain product!"

5.2 The Gromov–Witten potential and the quantum product

5.2.1 Identifying the invariants. We want to construct the generating function for the Gromov–Witten invariants. The Gromov–Witten invariants depend on the degree d and the input classes $\gamma_1, \dots, \gamma_n$, which can vary freely in $A^*(\mathbb{P}^r)$. By linearity of the Gromov–Witten invariants we can get a more manageable indexing set by allowing only input classes that belong to a basis for $A^*(\mathbb{P}^r)$. We will always use the standard basis

$$\{h^0, h^1, \dots, h^{r-1}, h^r\},$$

where h^0 is the fundamental class, h^1 is the hyperplane class, and h^r is the class of a point.

So now the possible input classes $(h^0)^{\bullet a_0}(h^1)^{\bullet a_1} \cdots (h^r)^{\bullet a_r}$ are parametrized by the index variables $\mathbf{a} = (a_0, \dots, a_r) \in \mathbb{N}^{r+1}$. Observe (once again) that the order of the input factors is immaterial, so we only need to bother about *how many* factors there are of each h^i, not about their position.

5.2.2 Collecting the degrees. We can get rid of the parameter d simply by defining "collected Gromov–Witten invariants"

$$I(\gamma_1 \cdots \gamma_n) := \sum_{d=0}^{\infty} I_d(\gamma_1 \cdots \gamma_n).$$

For this to make sense we must argue that only finitely many terms are nonzero; this comes about for dimension reasons: by the first reduction step we can assume all the input classes are homogeneous, say of codimensions $c_1, \dots, c_n$, so the total codimension is $\sum c_i$. On the other hand, the space $\overline{M}_{0,n}(\mathbb{P}^r, d)$ where the Gromov–Witten invariant is computed is of dimension $rd + r + d + n - 3$, so we get a contribution only when d is such that these two numbers are equal. In other words, only for

$$d = \frac{\sum c_i - r - n + 3}{r + 1}$$

can we have $I_d(\gamma_1 \cdots \gamma_n) \neq 0$. So in fact there is at most one term in the sum, and thus conversely we can recover I_d if we know I.

5.2.3 The Gromov–Witten potential. So now the Gromov–Witten invariants $I\big((h^0)^{\bullet a_0}(h^1)^{\bullet a_1} \cdots (h^r)^{\bullet a_r}\big)$ are arranged in an array of size $\mathbb{N}^{r+1}$, indexed by $\mathbf{a} := (a_0, \ldots, a_r)$. The Gromov–Witten potential is the generating function for these numbers. Introduce formal variables $\mathbf{x} = (x_0, \ldots, x_r)$ corresponding to the indices $\mathbf{a} = (a_0, \ldots, a_r)$ and form the generating function

$$\Phi(x_0, \ldots, x_r) := \sum_{a_0, \ldots, a_r} \frac{x_0^{a_0} \cdots x_r^{a_r}}{a_0! \cdots a_r!} I\big((h^0)^{\bullet a_0}(h^1)^{\bullet a_1} \cdots (h^r)^{\bullet a_r}\big). \qquad (5.2.3.1)$$

It is practical to introduce multi-index notation. With $\mathbf{x} = (x_0, \ldots, x_r)$ and $\mathbf{a} = (a_0, \ldots, a_r)$, put

$$\mathbf{x}^{\mathbf{a}} = x_0^{a_0} \cdots x_r^{a_r} \qquad \text{and} \qquad \mathbf{a}! = a_0! \cdots a_r!\,.$$

We also agree on the notation $\mathbf{h}^{\mathbf{a}} = (h^0)^{\bullet a_0}(h^1)^{\bullet a_1} \cdots (h^r)^{\bullet a_r}$. The benefit of this compact notation is that things look as they do in the univariate case: we can write

$$\Phi(\mathbf{x}) = \sum_{\mathbf{a}} \frac{\mathbf{x}^{\mathbf{a}}}{\mathbf{a}!} I(\mathbf{h}^{\mathbf{a}}).$$

Whenever we sum over $\mathbf{a}$ like this, it is understood that $\mathbf{a}$ runs over $\mathbb{N}^{r+1}$.

5.2.4 Derivatives of the Gromov–Witten potential. It follows from the derivative rule 5.1.2 that $\Phi_i := \frac{\partial}{\partial x_i}\Phi$ is the generating function for the Gromov–Witten invariants with an extra input class h^i. Precisely,

$$\Phi_i = \sum_{\mathbf{a}} \frac{\mathbf{x}^{\mathbf{a}}}{\mathbf{a}!} I(\mathbf{h}^{\mathbf{a}} \cdot h^i).$$

To check this you may want to revert to the expanded notation of (5.2.3.1). In particular we get the following expression for the third partial derivatives of Φ:

$$\Phi_{ijk} = \sum_{\mathbf{a}} \frac{\mathbf{x}^{\mathbf{a}}}{\mathbf{a}!} I(\mathbf{h}^{\mathbf{a}} \cdot h^i \cdot h^i \cdot h^k).$$

5.2.5 A more intrinsic description. The following point of view will be very useful in what follows. It is yet another formal manipulation. Consider the variables $\mathbf{x} = (x_0, \ldots, x_r)$ as generic coordinates on $A^*(\mathbb{P}^r)$ with respect to the basis $h^0, \ldots, h^r$. In other words, a generic element $\gamma \in A^*(\mathbb{P}^r)$ is written

$$\gamma = \sum_{i=0}^{r} x_i h^i.$$

Now we can hide all the formal variables, writing

$$\Phi = I(\exp(\gamma)) = \sum_{n=0}^{\infty} \frac{1}{n!} I(\gamma^{\bullet n}), \tag{5.2.5.1}$$

which you can think of as a coordinate-free definition of the potential. (Again, the bullet in the exponent of γ is just to remind us that we are not talking about the cup product of n classes in $A^*(X)$, but that there are n classes as input for the Gromov–Witten invariant.)

To establish this claim—or rather, to make sense of these formal expressions—just write out the definition of the exponential series. The right-hand equation is just the expansion $\exp(\gamma) = \sum_{n\geq 0} \frac{\gamma^n}{n!}$ combined with the linearity of I. On the other hand, writing out in coordinates we get

$$\begin{aligned}
\exp(\gamma) &= \exp\Big(\sum_{i=0}^{r} x_i h^i\Big) \\
&= \prod_{i=0}^{r} \exp(x_i h^i) \\
&= \prod_{i=0}^{r} \Big(\sum_{a_i=0}^{\infty} \frac{x_i^{a_i}}{a_i!} (h^i)^{a_i} \Big) \\
&= \sum_{\mathbf{a}} \frac{\mathbf{x}^{\mathbf{a}}}{\mathbf{a}!} \mathbf{h}^{\mathbf{a}}.
\end{aligned}$$

Now take I of these expressions and invoke linearity to retrieve (5.2.5.1).

One concrete advantage of the interpretation (5.2.5.1), in addition to being convenient compact notation, is that we now sum over the number $n = a_0 + \cdots + a_r$, the total number of marks used in the definition of the Gromov–Witten invariant.

5.2.6 Cohomology of $\mathbb{P}^r$ and the classical product. The starting point is the Chow ring $A^*(\mathbb{P}^r)$ with its "classical product" $\cup$. We will always work with the basis $\{h^0, h^1, \ldots, h^{r-1}, h^r\}$. It is immediate that we have the relations

$$\int_{\mathbb{P}^r} h^i \cup h^j = \begin{cases} 0 & \text{when } i + j \neq r \\ 1 & \text{when } i + j = r. \end{cases} \tag{5.2.6.1}$$

More generally, we have $h^i \cup h^j = h^{i+j}$. This equation can also be written in terms of some Gromov–Witten invariants:

$$h^i \cup h^j = \sum_{e+f=r} I_0(h^i \cdot h^j \cdot h^e)\, h^f .$$

To see this, note first that $I_0(h^i \cdot h^j \cdot h^e) = \int_{\mathbb{P}^r} h^i \cup h^j \cup h^e$; cf. Lemma 4.2.1. Next use (5.2.6.1).

We are now going to introduce a new kind of product, the *quantum product.* Instead of using only those few degree-0 invariants, the quantum product uses *all* the Gromov–Witten invariants, i.e., the whole Gromov–Witten potential, and in this way it encodes enumerative information.

Definition. The *quantum product* $*$ is defined by

$$h^i * h^j := \sum_{e+f=r} \Phi_{ije}\, h^f .$$

Whenever we write a sum indexed like this, it is understood that e and f are nonnegative integers, of course.

The right-hand side is an element in $A^*(\mathbb{P}^r) \otimes_{\mathbb{Z}} \mathbb{Q}[[\mathbf{x}]]$. Extending $\mathbb{Q}[[\mathbf{x}]]$-linearly defines the product in all of $A^*(\mathbb{P}^r) \otimes_{\mathbb{Z}} \mathbb{Q}[[\mathbf{x}]]$, the *quantum cohomology.* In general, when a multiplication map is defined in coordinates like this, the coefficients are called structure constants, so in this terminology we can say that the third derivatives of the Gromov–Witten potential are structure constants for the quantum multiplication.

Since obviously Φ_{ijk} is symmetric in the indices, we have the following:

5.2.7 Lemma. *The quantum product is commutative.* □

5.2.8 Remark. If one of the three indices of Φ_{ijk} is zero, say $i = 0$, then

$$\begin{aligned}
\Phi_{0jk} &= \sum_{n=0}^{\infty} \frac{1}{n!} \sum_{d \geq 0} I_d(\gamma^{\bullet n} \cdot h^0 \cdot h^j \cdot h^k) \\
&= I_0(h^0 \cdot h^j \cdot h^k) \\
&= \int_{\mathbb{P}^r} h^j \cup h^k ,
\end{aligned}$$

because the only Gromov–Witten invariants including a fundamental class are those of degree zero with 3 marks (according to 4.2.1).

5.2.9 Lemma. *The fundamental class h^0 is the identity for $*$.*

Proof. Using the previous observation we can write

$$h^0 * h^i = \sum_{e+f=r} (\textstyle\int h^i \cup h^e) h^f = h^i. \qquad \square$$

5.3 Associativity

The central result about the quantum product is its associativity.

5.3.1 Theorem. *The quantum product is associative. That is,*

$$(h^i * h^j) * h^k = h^i * (h^j * h^k).$$

Proof. The result is not much more than a formal consequence of the linear equivalence among the boundary divisors, $D(p_1 p_2 | p_3 p_4) \equiv D(p_2 p_3 | p_1 p_4)$ (cf. 2.7.5.1), together with the splitting lemma 4.3.2. The only difficulty is the notation, which tends to be a bit messy.

Let us first expand the two sides of the associativity relation to see what it actually means. On the left-hand side we find

$$(h^i * h^j) * h^k = \Big(\sum_{e+f=r} \Phi_{ije} h^f\Big) * h^k = \sum_{e+f=r} \sum_{l+m=r} \Phi_{ije} \Phi_{fkl} h^m.$$

Expanding the right-hand side in the same way we see that the associativity is given by

$$\sum_{e+f=r} \sum_{l+m=r} \Phi_{ije} \Phi_{fkl} h^m = \sum_{e+f=r} \sum_{l+m=r} \Phi_{jke} \Phi_{fil} h^m,$$

and since the h^m are linearly independent, this is equivalent to having

$$\sum_{e+f=r} \Phi_{ije} \Phi_{fkl} = \sum_{e+f=r} \Phi_{jke} \Phi_{fil} \qquad \text{for every } i, j, k, l.$$

These differential equations are called the *WDVV equations*, after Witten, Dijkgraaf, Verlinde, and Verlinde. There are two things to note about this equation. First, each side is a product of two generating functions, so according to the general principle 5.1.4, it corresponds to a count of "two-part things," in the present situation this means that we are counting two-component stable maps (i.e., points on the boundary of the moduli space). The second thing to note in the WDVV equation is the permutation of the four indices i, j, k, l, which is also characteristic for the fundamental linear equivalence of boundary divisors.

Now let us extract a recursive relation from the differential equation and see in concrete terms what the two remarks amount to. The series $\Phi_{ije} = \sum_n \frac{1}{n!} I(\gamma^{\bullet n} \cdot$

$h^i \cdot h^j \cdot h^e)$ is the generating function for the invariants $I(\gamma^{\bullet n} \cdot h^i \cdot h^j \cdot h^e)$ (but note that the formal variables are hidden in these invariants, instead of appearing on top of the factorials as usual). So by the product rule 5.1.4 we see that the left-hand side of the WDVV equation is the generating function for the invariants

$$\sum_{e+f=r} \sum_{n_A+n_B=n} \frac{n!}{n_A! n_B!} I(\gamma^{\bullet n_A} \cdot h^i \cdot h^j \cdot h^e)\, I(\gamma^{\bullet n_B} \cdot h^f \cdot h^k \cdot h^l).$$

So the associativity equations (WDVV equations) are equivalent to this:

$$\begin{aligned} &\sum_{e+f=r} \sum_{n_A+n_B=n} \frac{n!}{n_A! n_B!} I(\gamma^{\bullet n_A} \cdot h^i \cdot h^j \cdot h^e)\, I(\gamma^{\bullet n_B} \cdot h^f \cdot h^k \cdot h^l) \\ &\quad = \sum_{e+f=r} \sum_{n_A+n_B=n} \frac{n!}{n_A! n_B!} I(\gamma^{\bullet n_A} \cdot h^j \cdot h^k \cdot h^e)\, I(\gamma^{\bullet n_B} \cdot h^f \cdot h^i \cdot h^l) \end{aligned} \tag{5.3.1.1}$$

We will now show how this equation is a direct consequence of the linear equivalence 2.7.5.1. Fix d and n arbitrary and consider the space $\overline{M}_{0,n+4}(\mathbb{P}^r, d)$ with four important marks p_1, p_2, p_3, p_4, and further n marks we will not need to distinguish. Start out with the fundamental linear equivalence of boundary divisors

$$D(p_1 p_2 | p_3 p_4) \equiv D(p_2 p_3 | p_1 p_4).$$

Consider the four classes h^i, h^j, h^k, h^l and take their pullback via the evaluation maps corresponding to the four important marks. Consider further the pullbacks of n copies of γ along the evaluation maps corresponding to the remaining n marks. As in the previous chapter, we use the symbol $\underline{\nu}^*(\underline{\gamma})$ to denote the cup product of these n classes.

Now we integrate the product of these classes over the two equivalent boundary divisors, obtaining the identity

$$\begin{aligned} &\int_{D(p_1 p_2 | p_3 p_4)} \underline{\nu}^*(\underline{\gamma}) \cup \nu_1^*(h^i) \cup \nu_2^*(h^j) \cup \nu_3^*(h^k) \cup \nu_4^*(h^l) \\ &\quad = \int_{D(p_2 p_3 | p_1 p_4)} \underline{\nu}^*(\underline{\gamma}) \cup \nu_1^*(h^i) \cup \nu_2^*(h^j) \cup \nu_3^*(h^k) \cup \nu_4^*(h^l). \end{aligned}$$

Let us expand the left-hand side. The divisor $D(p_1 p_2 | p_3 p_4)$ is made up of several components, corresponding to all possible ways of distributing the n unspecified marks and the degree on the two twigs. In other words, it is the sum of the $\frac{n!}{n_A! n_B!}$ divisors of type indicated by this figure:

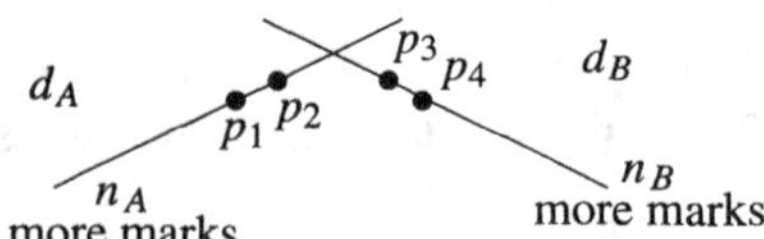

Note that since the n classes γ are equal, all these components of $D(p_1, p_2|p_3, p_4)$ give the same contribution. To each of the components we apply the splitting lemma (in fact, its Corollary 4.3.3) to obtain the following expression on the left-hand side of the equivalence:

$$\sum_{\substack{d_A+d_B=d \\ n_A+n_B=n}} \frac{n!}{n_A! n_B!} \left(\sum_{e+f=r} I_{d_A}(\gamma^{\bullet n_A} \cdot h^i \cdot h^j \cdot h^e)\, I_{d_B}(\gamma^{\bullet n_B} \cdot h^k \cdot h^l \cdot h^f) \right).$$

Now summing over all d we arrive exactly at the left-hand side of (5.3.1.1). The same arguments applied to the right-hand side establish the equation, and thus the associativity. □

5.4 Kontsevich's formula via quantum cohomology

5.4.1 The classical potential and the quantum potential. In order to extract enumerative information from the associativity relation, it is convenient to decompose the potential into a degree-zero part and a positive-degree part:

$$\Phi = \Phi^{\mathrm{cl}} + \Gamma.$$

The enumerative information (concerning honest curves, not contracted to points) is in the $(d > 0)$-part,

$$\Gamma = \sum_{n=0}^{\infty} \frac{1}{n!}\, I_+(\gamma^{\bullet n}),$$

where $I_+ = \sum_{d>0} I$. But let us first take a look at $\Phi^{\mathrm{cl}} = \sum_{n=0}^{\infty} \frac{1}{n!} I_0(\gamma^{\bullet n})$, which is called the *classical potential*. Recall from 4.2.1 that the only Gromov–Witten invariants in degree zero are those with precisely three marks, and we have $I_0(\gamma_1 \cdot \gamma_2 \cdot \gamma_3) = \int_{\mathbb{P}^r} \gamma_1 \cup \gamma_2 \cup \gamma_3$. Hence Φ^{cl} is a cubic polynomial

$$\Phi^{\mathrm{cl}} = \sum_{i,j,k} \frac{x_i x_j x_k}{3!} I_0(h^i \cdot h^j \cdot h^k).$$

(The factorial 3! takes care of repetitions due to the symmetry of the situation.) Clearly $\Phi^{\mathrm{cl}}_{ijk} = I_0(h^i \cdot h^j \cdot h^k)$, so the third derivatives of the classical potential are the structure constants for the classical product,

$$h^i \cup h^j = \sum_{e+f=r} \Phi^{\mathrm{cl}}_{ije}\, h^f,$$

just as the third derivatives of the full potential are structure constants for the quantum product.

Now we can decompose the quantum product into the classical part and the quantum part:

$$\begin{aligned} h^i * h^j &= \sum_{e+f=r} \big(I_0(h^i \cdot h^j \cdot h^e) + \Gamma_{ije}\big)h^f \\ &= (h^i \cup h^j) + \sum_{e+f=r} \Gamma_{ije} h^f . \end{aligned}$$

5.4.2 The quantum product for $\mathbb{P}^2$. We now restrict our attention to the case of $\mathbb{P}^2$. Here we have only three classes to treat: h^0, h^1, and h^2. Let us write down the multiplication table of the quantum product explicitly. To this end, note first that $\Gamma_{ije} = 0$ whenever one of the three indices i, j, e vanishes; this is a consequence of the fact that the presence of a fundamental class in a Gromov–Witten invariant makes it vanish, except in degree 0 (cf. 4.2.3). Writing the product as classical part plus quantum part, we get

$$\begin{aligned} h^1 * h^1 &= h^2 + \Gamma_{111}h^1 + \Gamma_{112}h^0 \\ h^1 * h^2 &= \Gamma_{121}h^1 + \Gamma_{122}h^0 \\ h^2 * h^2 &= \Gamma_{221}h^1 + \Gamma_{222}h^0 . \end{aligned}$$

Let us write down what the associativity relation says. There are only two nontrivial cases: $(h^1 * h^1) * h^2 = h^1 * (h^1 * h^2)$ and $(h^1 * h^2) * h^2 = h^1 * (h^2 * h^2)$. Let us work out the first one: on the one hand,

$$(h^1 * h^1) * h^2 = \Gamma_{221}h^1 + \Gamma_{222}h^0 + \Gamma_{111}(\Gamma_{121}h^1 + \Gamma_{122}h^0) + \Gamma_{112}h^2,$$

while on the other hand,

$$h^1 * (h^1 * h^2) = \Gamma_{121}(h^2 + \Gamma_{111}h^1 + \Gamma_{112}h^0) + \Gamma_{122}h^1 .$$

Equating the h^0-terms, we obtain the relation

$$\Gamma_{222} + \Gamma_{111}\Gamma_{122} = \Gamma_{112}\Gamma_{112}. \tag{5.4.2.1}$$

(The second case turns out to give the same relation, this time as the coefficients of h^1.)

We will now translate this differential equation into a recursion for the numbers generated by the involved series, so let us first make explicit in which sense these series are generating functions for which numbers. We have

$$\begin{aligned}\Gamma_{ijk} &= \sum_{n=0}^{\infty} \frac{1}{n!} I_+(\gamma^{\bullet n} \cdot h^i \cdot h^j \cdot h^k) \\ &= \sum_{\mathbf{a}} \frac{\mathbf{x}^{\mathbf{a}}}{\mathbf{a}!} I_+(\mathbf{h}^{\mathbf{a}} \cdot h^i \cdot h^j \cdot h^k).\end{aligned}$$

Here we revive the notation with all the formal variables (recall that $\gamma = x_0 h^0 + x_1 h^1 + x_2 h^2$), because now it is opportune to set $x_0 = x_1 = 0$. What this means is that we consider the special case $\gamma = x_2 h^2$. It is clear that the equation continues to hold when we do this substitution. What is interesting to note is that in fact we do not throw away any information in this way. Indeed, we have already observed that I_+ is zero whenever there is a factor h^0. This shows that the only value of a_0 giving a contribution is $a_0 = 0$, so actually Γ is independent of x_0. It is trickier to set $x_1 = 0$, but Lemma 4.2.4 shows that the Gromov–Witten invariants with a factor h^1 are completely determined by those without such a factor. So in fact we do not lose any information by making this reduction.

For simplicity call $x_2 = x$, so

$$\Gamma_{ijk} = \sum_{n=0}^{\infty} \frac{x^n}{n!} \, I_+((h^2)^{\bullet n} \cdot h^i \cdot h^j \cdot h^k)$$

is the generating function for the numbers $I_+((h^2)^{\bullet n} \cdot h^i \cdot h^j \cdot h^k)$. Hence, by the product rule 5.1.4, the differential equation (5.4.2.1) corresponds to this recursion:

$$\begin{aligned}I_+((h^2)^{\bullet n} h^2 h^2 h^2) \; + \sum_{n_A+n_B=n} & \tfrac{n!}{n_A! n_B!} \, I_+((h^2)^{\bullet n_A} h^1 h^1 h^1) \; I_+((h^2)^{\bullet n_B} h^1 h^2 h^2) \\ = \sum_{n_A+n_B=n} & \tfrac{n!}{n_A! n_B!} \, I_+((h^2)^{\bullet n_A} h^1 h^1 h^2) \; I_+((h^2)^{\bullet n_B} h^1 h^1 h^2).\end{aligned}$$

It remains to interpret the numbers $I_+((h^2)^{\bullet n} \cdot h^i \cdot h^j \cdot h^k)$. Each is a sum over $d > 0$, but only compatible values of d and n give contribution. We have $n + 3$ marks, and thus our space is $\overline{M}_{0,n+3}(\mathbb{P}^2, d)$, whose dimension is $3d+2+n$. On the other hand, the sum of the codimensions of the classes is $\sum \operatorname{codim} = 2n+i+j+k$. Equating these two numbers we find that only the case

$$n = 3d + 2 - i - j - k \tag{5.4.2.2}$$

gives any contribution. We can substitute this into the five Gromov–Witten invariants of the formula. Next, we can use Lemma 4.2.4 to move the h^1-factors outside I_d, where they become a factor d instead. Finally, recall from 4.1.6 that $I_d\big((h^2)^{\bullet 3d-1}\big) = N_d$. As illustration, let us perform these three steps on $I_+((h^2)^{\bullet n_A} \cdot h^1 \cdot h^1 \cdot h^2)$ — here the selection rule (5.4.2.2) reads $n_A = 3d_A - 2$.

$$\begin{aligned} I_+((h^2)^{\bullet n_A} \cdot h^1 \cdot h^1 \cdot h^2) &= I_{d_A}((h^2)^{\bullet 3d_A-2} \cdot h^1 \cdot h^1 \cdot h^2) \\ &= d_A^2\, I_{d_A}((h^2)^{\bullet 3d_A-1}) \\ &= d_A^2\, N_{d_A}. \end{aligned}$$

Doing this for each of the five Gromov–Witten invariants in the recursion relation, we arrive at

$$\begin{aligned} N_d \quad &+ \textstyle\sum_{d_A+d_B=d} \frac{(3d-4)!}{(3d_A-1)!(3d_B-3)!}\, d_A^3\, N_{d_A}\, d_B\, N_{d_B} \\ &= \textstyle\sum_{d_A+d_B=d} \frac{(3d-4)!}{(3d_A-2)!(3d_B-2)!}\, d_A^2\, N_{d_A}\, d_B^2\, N_{d_B}, \end{aligned}$$

which is precisely Kontsevich's formula.

5.5 Generalizations and references

5.5.1 More general smooth projective varieties. In this chapter a few simplifications were possible since we considered only the case of $\mathbb{P}^r$, but the theory is essentially the same for any projective homogeneous variety X: If $T_0, \ldots, T_r$ is a basis for $A^*(X)$, the intersection pairing matrix (g_{ij}) defined as

$$g_{ij} = \int_X T_i \cup T_j$$

comes into play. Since the intersection pairing is nondegenerate, the matrix (g_{ij}) is invertible; let (g^{ij}) be its inverse. For $\mathbb{P}^2$ we have

$$(g_{ij}) = (g^{ij}) = \begin{pmatrix} 0 & 0 & 1 \\ 0 & 1 & 0 \\ 1 & 0 & 0 \end{pmatrix}.$$

In this setting, the quantum product is

$$T_i * T_j = \sum_{e,f} \Phi_{ije} g^{ef} T_f.$$

In general, all the sums we had of type $\sum\limits_{e+f=r}$ become sums of type $\sum\limits_{e,f} g^{ef}$.

For general smooth projective varieties, the Gromov–Witten potential still makes sense, but with the lack of a selection-rule argument like 5.2.2, it is necessary to link the degree to yet another formal variable. Also, one has to use the cohomology ring $H^*(X)$ instead of the Chow ring $A^*(X)$, and since the classes of odd degree anticommute, there are then a lot of signs to keep track of. (It should be mentioned that enumerative interpretations are not possible in this general case.)

5.5.2 The small quantum cohomology ring. A variation of quantum cohomology that provides substantial simplification is the notion of the *small* quantum cohomology ring (cf. FP-NOTES, Section 10); historically, this is actually the original quantum cohomology; cf. Witten [86] and the references given there. Instead of using the full third derivatives as structure constants, all the variables except those corresponding to divisor classes are set to zero.

To get a glimpse of how that changes the scenario, let us briefly look at $\mathbb{P}^r$: the only divisor class is h^1, so the relevant variable is $x_1 = x$. After zeroing the others, we get

$$\begin{aligned}\Phi_{ijk}(x) &= \sum_{n=0}^{\infty} \frac{x^n}{n!} \sum_{d \geq 0} I_d(h^{\bullet n} \cdot h^i \cdot h^j \cdot h^k) \\ &= \sum_{n=0}^{\infty} \frac{x^n}{n!} \sum_{d \geq 0} d^n \cdot I_d(h^i \cdot h^j \cdot h^k)\end{aligned}$$

via Lemma 4.2.4. Hence, only invariants with three marks are involved. Note that to get any contribution at all from $I_d(h^i \cdot h^j \cdot h^k)$ we need $rd + r + d = i + j + k$, which is possible only for $d = 0$ (the classical part) and for $d = 1$ (which is then the quantum part). Setting $q := \exp(x) = \sum_n \frac{x^n}{n!}$ we get

$$\Phi_{ijk} = I_0(h^i \cdot h^j \cdot h^k) + q \cdot I_1(h^i \cdot h^j \cdot h^k).$$

The following description of the small quantum product follows readily:

$$h^i * h^j = \begin{cases} h^{i+j} & \text{for } i + j \leq r, \\ q\, h^{i+j-r-1} & \text{for } r < i + j \leq 2r. \end{cases}$$

So while the classical ring is $A^*(\mathbb{P}^r) \simeq \mathbb{Z}[h]/(h^{r+1})$, the small quantum ring is isomorphic to

$$\mathbb{Z}[h, q]/(h^{r+1} - q).$$

For $\mathbb{P}^r$, the small quantum product does not encode any interesting enumerative information. However, for more general varieties, even the 3-pointed Gromov–Witten invariants become interesting. See for example Crauder–Miranda [14] for rational surfaces, Beauville [5] for certain complete intersections, or Qin–Ruan [70] for projective bundles.

A great deal of interest in small quantum cohomology comes from combinatorics: there is a highly developed theory for Grassmannians and flag manifolds, which has led to interesting generalizations of classical combinatorics; one feature of small quantum cohomology is that you can actually compute a lot! To mention a few papers, Bertram [9] and Bertram–Ciocan-Fontanine–Fulton [10] do Grassmannians (quantum Giambelli, quantum Pieri, quantum Littlewood–Richardson), and the recent Fulton–Woodward [30] does general G/P.

5.5.3 Tangency quantum cohomology. From the viewpoint of enumerative geometry, an interesting generalization of quantum cohomology is that of tangency quantum cohomology; cf. Kock [54]. It is a sort of quantum product which encodes not only incidence conditions (Gromov–Witten invariants) but also tangency conditions (certain gravitational descendants, briefly mentioned in 4.5.5). This construction involves a more comprehensive potential, with new variables $\mathbf{y} = (y_0, \dots, y_r)$ for the tangency conditions, and also a more comprehensive "metric," which (unlike the constants g_{ij}) depends on $\mathbf{y}$. For $\mathbb{P}^2$, the tangency quantum product had previously been constructed by Ernström and Kennedy [24] exploiting a space of stable lifts (briefly mentioned in 4.5.1).

5.5.4 Frobenius manifolds. To give an idea of how quantum cohomology is situated in a more general context, we conclude the exposition with an exercise in Riemannian geometry(!) (see for example do Carmo [18] for the definitions).

Let X be a projective homogeneous variety (a Grassmannian, say), and consider the vector space $V = H^*(X, \mathbb{C})$ as a differentiable manifold. Let $T_0, \dots, T_r$ be a basis (formed by the Schubert cycles, say, if X is a Grassmannian), and let $\partial_0, \dots, \partial_r$ be the corresponding vector fields. The (g_{ij}) define a metric on V by $\langle\, \partial_i \,|\, \partial_j \,\rangle := g_{ij}$, called the *Poincaré metric*. Define a (formal) connection ∇ by its Christoffel symbols $A_{ij}^f := \sum_e \Phi_{ije} g^{ef}$, that is,

$$\nabla_{\partial_i} \partial_j = \sum_f A_{ij}^f \partial_f = \sum_{e,f} \Phi_{ije} g^{ef} \partial_f .$$

Recall that the curvature of a connection A_{ij}^f is given in coordinates by

$$R(\partial_i, \partial_j)\partial_k = \sum_m R_{ijk}^m \partial_m ,$$

where (cf. [18], p. 93)

$$R_{ijk}^m = \sum_f A_{ik}^f A_{fj}^m - \sum_f A_{jk}^f A_{fi}^m + \partial_j A_{ik}^m - \partial_i A_{jk}^m .$$

A connection is called *flat* if its curvature is identically zero.

Now we claim: *The connection ∇ defined above is flat if and only if the quantum product is associative.*

Let us see: in the expression of R^m_{ijk}, the last two terms cancel out thanks to the observation that $\Phi_{ijk} = \partial_i\partial_j\partial_k\Phi$, and that the order of the partial derivations is irrelevant. Accordingly, $\partial_j A^m_{ik} = \sum_l \partial_j \Phi_{ikl} g^{lm} = \sum_l \partial_i \Phi_{jkl} g^{lm} = \partial_i A^m_{jk}$. Now let us expand the first two terms of R^m_{ijk}:

$$\sum_f A^f_{ik} A^m_{fj} - \sum_f A^f_{jk} A^m_{fi} = \sum_{e,f,l} \Phi_{ike} g^{ef} \Phi_{fjl} g^{lm} - \sum_{e,f,l} \Phi_{jke} g^{ef} \Phi_{fil} g^{lm}.$$

Since (g^{lm}) is invertible, the vanishing of this expression is equivalent to having for all l the identity

$$\sum_{e,f} \Phi_{ike} g^{ef} \Phi_{fjl} - \sum_{e,f} \Phi_{jke} g^{ef} \Phi_{fil} = 0,$$

which is nothing but the associativity relation.

This formalism is due to Dubrovin [19] and has been explored since the original paper of Kontsevich and Manin [58]. Thus quantum cohomology provides an important class of examples of the following general notion:

Definition. A (formal) *Frobenius manifold* is a Riemannian manifold (V, g) with a (formal) flat connection A^f_{ij}, satisfying the following integrability condition: there exists a "potential" Φ such that $A^f_{ij} = \sum_e \Phi_{ije} g^{ef}$.

Frobenius manifolds appear in other areas of mathematics, such as for example, integrable systems (see Dubrovin [19]) and singularity theory (see Hertling [45]).

5.5.5 CohFT and Frobenius manifolds. We saw in this chapter how the associativity (cf. 5.3.1) is a consequence of the Splitting Lemma 4.3.2. There is a generalization of this principle that we mention briefly. While the associativity (together with the existence of the potential) generalizes into the concept of a Frobenius manifold, the recursion lemma has as a generalization the CohFT structures (cf. 4.5.6). There is the following theorem (cf. Manin's book [61], Ch. III, Th. 4.3): *Having a CohFT structure on X is equivalent to having a Frobenius manifold structure on* $H^*(X, \mathbb{C})$ (in the sense that one can construct one structure from the other without loss of information).

5.5.6 Readings. Everyone should read (or have a look at) the epoch-making survey of Witten [86]; at least have a look at § 3 where the basic theory of quantum cohomology is outlined. Here "survey" does not only mean "review of research developments," but as much a *preview* of a decade of remarkable mathematics.

Exercises

1. Compute the classical potentials for $\mathbb{P}^1$, $\mathbb{P}^2$, and $\mathbb{P}^3$.

2. Show that the Gromov–Witten potential for $\mathbb{P}^1$ is

$$\Phi(x_0, x_1) = \frac{x_0^2 x_1}{2} + \exp(x_1).$$

Hint: you computed all the Gromov–Witten invariants of $\mathbb{P}^1$ in Exercise 6 on page 126.

3. Let Γ denote the quantum part of the Gromov–Witten potential for $\mathbb{P}^3$. Associativity of the quantum product for $\mathbb{P}^3$ is equivalent to the following system of equations:

$$\begin{aligned}
2\Gamma_{123} + \Gamma_{112}\Gamma_{122} &= \Gamma_{222} + \Gamma_{111}\Gamma_{222} \\
\Gamma_{133} + \Gamma_{113}\Gamma_{122} &= \Gamma_{223} + \Gamma_{111}\Gamma_{223} \\
\Gamma_{233} + \Gamma_{112}\Gamma_{133} + \Gamma_{111}\Gamma_{233} &= 2\Gamma_{113}\Gamma_{123} \\
\Gamma_{233} + \Gamma_{112}\Gamma_{223} &= \Gamma_{113}\Gamma_{222} \\
\Gamma_{333} + \Gamma_{122}\Gamma_{133} + \Gamma_{112}\Gamma_{233} &= \Gamma_{123}\Gamma_{123} + \Gamma_{113}\Gamma_{223} \\
2\Gamma_{123}\Gamma_{223} &= \Gamma_{133}\Gamma_{222} + \Gamma_{122}\Gamma_{233}.
\end{aligned}$$

Look carefully at the indices to see what the equations do:

(i) Show that the fourth equation expresses each Gromov–Witten invariant with at least two T_3-factors (point-classes) in terms of invariants of lower degree.

(ii) Show that the second equation expresses each invariant with one T_3-factor (and two T_2-factors) in terms of invariants with two T_3-factors, modulo lower degree invariants.

(iii) Show that the first equation expresses each invariant without T_3-factors in terms of invariants with one T_3-factor, modulo lower-degree invariants.

(iv) Show that those three equations together are sufficient to implement the recursion algorithm described in Chapter 4.

4. Let H denote the intersection ring $A^*(\mathbb{P}^r) \otimes \mathbb{Q}$. We work in the linear basis $\{h^0, h^1, \ldots, h^r\}$. Consider the formal exponential series in H,

$$\Phi := \sum_{\mathbf{a}} \frac{\mathbf{x}^{\mathbf{a}}}{\mathbf{a}!} \mathbf{h}^{\mathbf{a}} \in H[[\mathbf{x}]],$$

and its integral

$$\phi := \int_{\mathbb{P}^r} \Phi \in \mathbb{Q}[[\mathbf{x}]].$$

(i) Show that ϕ is equal to $\exp(x_0)$ times a polynomial in $x_1, \ldots, x_r$ of degree r, if we consider each variable x_i of degree i. Write explicit formulas for ϕ in the two cases $\mathbb{P}^2$ and $\mathbb{P}^3$.

(ii) Show that the classical potential of $\mathbb{P}^r$ is the degree-3 part of ϕ, if we consider each variable x_i to be of degree 1.

5. (Cf. [54].) Continuing the previous exercise, put $\Phi_i := \frac{\partial}{\partial x_i}\Phi = \Phi \cdot h^i$ and put $\phi_i := \frac{\partial}{\partial x_i}\phi = \int \Phi \cdot h^i = \int \Phi_i$. Make sure you understand all these equations.

(i) Since the h^i form a $\mathbb{Q}$-basis for H, they also form a $\mathbb{Q}[[\mathbf{x}]]$-basis for $H[[\mathbf{x}]]$. Use linear algebra to establish the formulas

$$\Phi = \sum_{e+f=r} \phi_e h^f \qquad \text{and} \qquad \Phi_i = \sum_{e+f=r} \phi_{ie} h^f.$$

(ii) Show that the matrix (ϕ_{ij}) is invertible over $\mathbb{Q}[[\mathbf{x}]]$, and conclude that the Φ_i form a $\mathbb{Q}[[\mathbf{x}]]$-basis for $H[[\mathbf{x}]]$.

(iii) Show that Φ is invertible in $H[[\mathbf{x}]]$ (with the usual product induced from H by $\mathbb{Q}[[\mathbf{x}]]$-linearity).

(iv) Define a new product on $H[[\mathbf{x}]]$ by the rule

$$h^i \star h^j := \sum_{e+f=r} \phi_{ije} h^f$$

in analogy with the quantum product. Show that Φ^{-1} is the identity element for this product.

(v) Consider two generic elements in H:

$$\gamma' = \sum x_i' h^i \qquad \text{and} \qquad \gamma'' = \sum x_i'' h^i.$$

Prove this sum formula:

$$\phi(\gamma' + \gamma'') = \sum_{e+f=r} \phi_e(\gamma')\phi_f(\gamma'').$$

Hint: consider γ' and γ'' as classes on two different copies of $\mathbb{P}^r$, consider the diagonal $\Delta \subset \mathbb{P}^r \times \mathbb{P}^r$, and compute the integral

$$\int_\Delta \exp(\gamma') \cup \exp(\gamma'')$$

in two ways: using the isomorphism $\Delta \simeq X$, and using the Künneth formula.

(vi) In the same spirit, establish the formula

$$\phi_{ijk\ell}(2\gamma) = \sum_{e+f=r} \phi_{ije}\phi_{fk\ell}$$

and conclude that the product $\star$ is associative.

6. The "quantum corrections" Γ_{ijk} to the standard cup product do not constitute a deformation of it; setting $\mathbf{x} = \mathbf{0}$ does not give back the classical product:

   (i) Show that

$$\Gamma_{ijk}(\mathbf{0}) = \begin{cases} 1 & \text{for } i + j + k = 2r + 1 \\ 0 & \text{otherwise}. \end{cases}$$

   (ii) Show that in setting $\mathbf{x} = \mathbf{0}$ in the quantum product, the resulting product gives a ring structure on A isomorphic to

$$\mathbb{Z}[h]/(h^{r+1} - 1).$$

   (This is just the $q = 1$ case of small quantum cohomology (cf. 5.5.2).

Quantum cohomology of $\mathbb{P}^1 \times \mathbb{P}^1$. See the exercises to Chapter 3 for some background. We employ the following basis: T_0 is the fundamental class; T_1 is the class of the horizontal rule, and T_2 the class of the vertical rule; T_3 is the class of a point.

7. Write down the classical potential for $\mathbb{P}^1 \times \mathbb{P}^1$.

8. Let Γ denote the quantum part of the Gromov–Witten potential for $\mathbb{P}^1 \times \mathbb{P}^1$. Show that associativity of the quantum product of $\mathbb{P}^1 \times \mathbb{P}^1$ implies the equation

$$\Gamma_{333} + \Gamma_{112}\Gamma_{233} + \Gamma_{122}\Gamma_{133} = \Gamma_{123}\Gamma_{123} + \Gamma_{113}\Gamma_{223}$$

   and show that this is precisely the differential equation corresponding to Kontsevich's formula for $\mathbb{P}^1 \times \mathbb{P}^1$; cf. Exercise 9 on page 109.

9. Show that the small quantum cohomology ring of $\mathbb{P}^1 \times \mathbb{P}^1$ is isomorphic to

$$\mathbb{Z}[h, v, q_h, q_v]/(h^2 - q_v, v^2 - q_h).$$

Quantum cohomology of $G(2,4)$*, the Grassmannian of lines in* $\mathbb{P}^3$. We employ the standard basis; cf. page 127.

10. Recall that the multiplication table is

$\cup$	T_0	T_1	T_2	T_3	T_4	T_5
T_0	T_0	T_1	T_2	T_3	T_4	T_5
T_1	T_1	T_2+T_3	T_4	T_4	T_5	0
T_2	T_2	T_4	T_5	0	0	0
T_3	T_3	T_4	0	T_5	0	0
T_4	T_4	T_5	0	0	0	0
T_5	T_5	0	0	0	0	0

Write down the classical potential, and show that conversely the classical potential determines the multiplication table, assuming that the top class T_5 has integral 1.

11. Show that among the Γ_{ijk}, the only ones that are nonzero after passing to the small quantum cohomology (i.e., after setting $x_0 = x_2 = x_3 = x_4 = x_5 = 0$; see 5.5.2) are

$$\Gamma_{145} = \Gamma_{235} = q, \qquad \Gamma_{555} = q^2.$$

Hint: most of the work was done in Exercise 9 on page 127.

12. Compute the multiplication table for the small quantum ring of G (cf. 5.5.2 for the definition), $q := \exp(x_1)$:

	T_0	T_1	T_2	T_3	T_4	T_5
T_0	T_0	T_1	T_2	T_3	T_4	T_5
T_1	T_1	T_2+T_3	T_4	T_4	T_5+qT_0	qT_1
T_2	T_2	T_4	T_5	qT_0	qT_1	qT_3
T_3	T_3	T_4	qT_0	T_5	qT_1	qT_2
T_4	T_4	T_5+qT_0	qT_1	qT_1	$q(T_2+T_3)$	qT_4
T_5	T_5	qT_1	qT_3	qT_2	qT_4	q^2T_0

Bibliography

FP-NOTES always refers to *Notes on Stable Maps and Quantum Cohomology* by W. Fulton and R. Pandharipande [29].

[1] Allen B. Altman and Steven L. Kleiman, *Introduction to Grothendieck Duality Theory*, Vol. 146, Lecture Notes in Mathematics, Springer-Verlag, 1970.

[2] Allen B. Altman and Steven L. Kleiman, Foundations of the theory of Fano schemes, *Compositio Math.* **34** (1977), 3–47.

[3] Michael F. Atiyah. Response to: Theoretical mathematics: toward a cultural synthesis of mathematics and theoretical physics, A. Jaffe and F. Quinn, *Bull. Amer. Math. Soc.* (N.S.) **30** (1994), 178–179.

[4] Arnaud Beauville, Surfaces algébriques complexes, *Astérisque* **54**, 1978.

[5] Arnaud Beauville, *Quantum cohomology of complete intersections*, preprint, (alg-geom/9501008).

[6] Kai Behrend, Gromov–Witten invariants in algebraic geometry, *Invent. Math.* **127** (1997), 601–617, (alg-geom/9601011).

[7] Kai Behrend and Barbara Fantechi, The intrinsic normal cone, *Invent. Math.* **128** (1997), 45–88, (alg-geom/9601010).

[8] Kai Behrend and Yuri I. Manin, Stacks of stable maps and Gromov–Witten invariants, *Duke. Math. J.* **85** (1996), 1–60, (alg-gcom/9506023).

[9] Aaron Bertram, Quantum Schubert calculus., *Adv. Math.* **128** (1997), 289–305, (alg-geom/9410024).

[10] Aaron Bertram, Ionuţ Ciocan-Fontanine, and William Fulton, Quantum multiplication of Schur polynomials, *J. Algebra* **219** (1999), 728–746, (alg-geom/9705024).

[11] Lucia Caporaso, Counting curves on surfaces: a guide to new techniques and results, In: *European Congress of Mathematics, Vol. I* (Budapest, 1996), pp. 136–149, Birkhäuser, Basel, 1998, (alg-geom/9611029).

[12] Lucia Caporaso and Joe Harris, Counting plane curves of any genus, *Invent. Math.* **131** (1998), 345–392, (alg-geom/9608025).

[13] David A. Cox and Sheldon Katz, *Mirror Symmetry and Algebraic Geometry*, American Mathematical Society, Providence, RI, 1999.

[14] Bruce Crauder and Rick Miranda, Quantum cohomology of rational surfaces, In: *The Moduli Space of Curves*, R. Dijkgraaf, C. Faber, and G. van der Geer, eds., Vol. 129, Progress in Mathematics, pp. 34–80, Birkhäuser, Boston, MA, 1995, (alg-geom/9410028).

[15] Pierre Deligne, *Resumé des premiers exposés de A. Grothendieck, SGA7, Exp. I*, Vol. 288, Lecture Notes in Mathematics, Springer-Verlag, New York, 1972, pp. 1–24.

[16] Pierre Deligne and David Mumford, On the irreducibility of the space of curves of given genus, *Publ. Math. I.H.E.S.* **36** (1969), 75–109.

[17] Philippe di Francesco and Claude Itzykson, Quantum intersection rings, In: *The moduli space of curves*, R. Dijkgraaf, C. Faber, and G. van der Geer, eds., Vol. 129, Progress in Mathematics, Birkhäuser, Boston, MA, 1995, pp. 81–148.

[18] Manfredo Perdigão do Carmo, *Geometria Riemanniana*, Instituto de Matemática Pura e Aplicada, Rio de Janeiro, 1988. English translation: *Riemannian Geometry*, Birkhäuser, Boston, 1992.

[19] Boris Dubrovin, Geometry of 2D topological field theories, In: *Integrable Systems and Quantum Groups*, Vol. 1620, Lecture Notes in Mathematics, pp. 120–348, Springer-Verlag, New York, 1996, (hep-th/9407018).

[20] Dan Edidin, Notes on the construction of the moduli space of curves, In: *Recent Progress in Intersection Theory (Bologna, 1997)*, pp. 85–113, Birkhäuser, Boston MA, 2000, (math.AG/9805101).

[21] Tohru Eguchi, Kentaro Hori, and Chuan-Sheng Xiong, Quantum cohomology and Virasoro algebra, *Phys. Lett. B* **402** (1997), 71–80, (hep-th/9703086).

[22] David Eisenbud and Joe Harris, *The Geometry of Schemes*, Vol. 197, Graduate Texts in Mathematics, Springer-Verlag, New York, 2000.

[23] Lars Ernström and Gary Kennedy, Recursive formulas for the characteristic numbers of rational plane curves, *J. Alg. Geom.* **7** (1998), 141–181, (alg-geom/9604019).

[24] Lars Ernström and Gary Kennedy, Contact cohomology of the projective plane, *Amer. J. Math.* **121** (1999), 73–96, (alg-geom/9703013).

[25] Eduardo Esteves, *Construção de espaços de moduli*, Instituto de Matemática Pura e Aplicada, Rio de Janeiro, 1997.

[26] Barbara Fantechi, *Stacks for everybody*, Available at http://www.cgtp.duke.edu/~drm/PCMI2001/fantechi-stacks.pdf.

[27] William Fulton, *Introduction to Intersection Theory in Algebraic Geometry*, Vol. 54, CBMS Reg. Conf. Series in Math., Amer. Math. Soc., Providence, RI, 1984.

[28] William Fulton, *Intersection Theory*, Springer-Verlag, New York, 1985.

[29] William Fulton and Rahul Pandharipande, *Notes on stable maps and quantum cohomology*, In: *Algebraic Geometry, Santa Cruz 1995*, J. Kollár, R. Lazarsfeld and D. Morrison, eds., Vol. 62, II of Proc. Symp. Pure. Math., pp. 45–96, 1997, (alg-geom/9608011).

[30] William Fulton and Christopher Woodward, On the quantum product of Schubert classes, *J. Alg. Geom.* **13** (2003), 641–661, (math.AG/0112183).

[31] Andreas Gathmann, Gromov–Witten invariants of blow-ups, *J. Alg. Geom.* **10** (2001), 399–432, (math.AG/9804043).

[32] Andreas Gathmann, Absolute and relative Gromov–Witten invariants of very ample hypersurfaces, *Duke. Math. J.* **115** (2002), 171–203, (math.AG/9908054).

[33] Letterio Gatto, *Intersection Theory over Moduli Spaces of Curves*, Vol. 61, Monografias de Matemática, IMPA, Rio de Janeiro, 2000.

[34] Ezra Getzler, Intersection theory on $\overline{M}_{1,4}$ and elliptic Gromov–Witten invariants, *J. Amer. Math. Soc.* **10** (1997), 973–998, (alg-geom/9612004).

[35] Ezra Getzler, The Virasoro conjecture for Gromov–Witten invariants, In: *Algebraic Geometry: Hirzebruch 70 (Warsaw, 1998)*, Vol. 241, Contemp. Math., pp. 147–176, Amer. Math. Soc., Providence, RI, 1999, (math.AG/9812026).

[36] Lothar Göttsche and Rahul Pandharipande, The quantum cohomology of blow-ups of $\mathbb{P}^2$ and enumerative geometry. *J. Diff. Geom.* **48** (1998), 61–90, (alg-geom/9611012).

[37] Tom Graber, Enumerative geometry of hyperelliptic plane curves, *J. Alg. Geom.* **10** (2001), 725–755, (alg-geom/9808084).

[38] Tom Graber, Joachim Kock, and Rahul Pandharipande, Descendant invariants and characteristic numbers, *Amer. J. Math* **124** (2002), 611–647, (math.AG/0102017).

[39] Tom Graber and Rahul Pandharipande, Localization of virtual classes, *Invent. Math.* **135** (1999), 487–518, (alg-geom/9708001).

[40] Mikhael Gromov, Pseudoholomorphic curves in symplectic manifolds, *Invent. Math.* **82** (1985), 307–347.

[41] Alexander Grothendieck, *Techniques de construction et théorèmes d'existence en géométrie algébrique. IV. Les schémas de Hilbert*, Séminaire Bourbaki, 13e année, 1960/61, exposé 221, 1961.

[42] Joe Harris, *Algebraic Geometry: A First Course*, Vol. 133, Graduate Texts in Mathematics, Springer-Verlag, New York, 1992.

[43] Joe Harris and Ian Morrison, *Moduli of Curves*, Vol. 187, Graduate Texts in Mathematics, Springer-Verlag, New York, 1998.

[44] Robin Hartshorne, *Algebraic Geometry*, Vol. 5, Graduate Texts in Mathematics, Springer-Verlag, New York, 1977.

[45] Claus Hertling, *Frobenius Manifolds and Moduli Spaces for Singularities*, Vol. 151, Cambridge Tracts in Mathematics, Cambridge University Press, Cambridge, 2002.

[46] Mikhail Kapranov, Veronese curves and Grothendieck–Knudsen moduli space $\overline{M}_{0,n}$, *J. Alg. Geom.* **2** (1993), 239–262.

[47] Sean Keel, Intersection theory of moduli spaces of stable n-pointed curves of genus zero, *Trans. Amer. Math. Soc.* **330** (1992), 545–574.

[48] Bumsig Kim and Rahul Pandharipande, The connectedness of the moduli space of maps to homogeneous spaces, In: *Symplectic Geometry and Mirror Symmetry (Seoul, 2000)*, pp. 187–201, World Sci. Publishing, River Edge, NJ, 2001, (math.AG/0003168).

[49] Steven L. Kleiman, The transversality of a general translate, *Compositio Math.* **28** (1974), 287–297.

[50] Steven L. Kleiman, Intersection theory and enumerative geometry: A decade in review, In: *Algebraic Geometry, Bowdoin 1985*, Vol. 46, Proc. Symp. Pure Math, pp. 321–370, 1987. With the collaboration of Anders Thorup on §3.

[51] Finn Knudsen, Projectivity of the moduli space of stable curves, II, *Math. Scand.* **52** (1983), 1225–1265.

[52] Joachim Kock, Recursion for twisted descendants and characteristic numbers of rational curves, preprint, math.AG/9902021.

[53] Joachim Kock, Characteristic numbers of rational curves with cusp or prescribed triple contact, *Math. Scand.* **92** (2003), 1–23, (math.AG/0102082).

[54] Joachim Kock, Tangency quantum cohomology, *Compositio Math.* **140** (2004), 165–178, (math.AG/0006148).

[55] Jánus Kollár, *Rational Curves on Algebraic Varieties*, Springer-Verlag, New York, 1996.

[56] Maxim Kontsevich, Intersection theory on the moduli space of curves and the matrix Airy function, *Comm. Math. Phys.* **147** (1992), 1–23.

[57] Maxim Kontsevich, Enumeration of rational curves via torus actions, In: *The moduli space of curves*, R. Dijkgraaf, C. Faber, and G. van der Geer, eds., Vol. 129, pp. 335–368, Progress in Mathematics, Birkhäuser, Boston, MA, 1995, (hep-th/9405035).

[58] Maxim Kontsevich and Yuri I. Manin, Gromov–Witten classes, quantum cohomology, and enumerative geometry, *Comm. Math. Phys.* **164** (1994), 525–562, (hep-th/9402147).

[59] Andrew Kresch, Farsta, computer program for computing Gromov–Witten invariants via associativity relations. Available at http://www.maths.warwick.ac.uk/~kresch/co/farsta.html.

[60] Saunders Mac Lane, *Categories for the Working Mathematician*, second edition, Vol. 5, Graduate Texts in Mathematics, Springer-Verlag, New York, 1998.

[61] Yuri I. Manin, Frobenius manifolds, quantum cohomology, and moduli spaces, AMS Colloquium Publications, Providence, RI, 1999.

[62] David Mumford, *Geometric Invariant Theory*, Springer-Verlag, Heidelberg, 1965. (David Mumford, John Fogarty, and Frances Kirwan, third enlarged edition, Springer-Verlag, Heidelberg, 1994.)

[63] David Mumford, *Abelian varieties*, Tata Institute of Fundamental Research Studies in Mathematics, No. 5. Published for the Tata Institute of Fundamental Research, Bombay, 1970.

[64] David Mumford, *Towards an enumerative geometry of the moduli space of curves*, In: *Arithmetic and Geometry, Volume II*, Vol. 36, Progress in Mathematics, pp. 271–328, Birkhäuser, Boston, MA, 1983.

[65] David Mumford, *The Red Book of Varieties and Schemes*, LNM 1358, Springer-Verlag, New York, 1988.

[66] Peter E. Newstead, *Introduction to Moduli Problems and Orbit Spaces*, Tata Institute Lecture Notes, Springer-Verlag, 1978.

[67] Rahul Pandharipande, Rational curves on hypersurfaces (after A. Givental), In: Séminaire Bourbaki 1997–1998, exposé 848, *Astérisque*, **252** (1998), 307–340, (math.AG/9806133).

[68] Rahul Pandharipande, A geometric construction of Getzler's relation, *Math. Ann.* **313** (1999), 715–729, (alg-geom/9705016).

[69] Rahul Pandharipande, Intersections of $\mathbb{Q}$-divisors on Kontsevich's moduli space $\overline{M}_{0,n}(\mathbb{P}^r, d)$ and enumerative geometry, *Trans. Amer. Math. Soc.* **351** (1999), 1481–1505, (alg-geom/9504004).

[70] Zhenbo Qin and Yongbin Ruan, Quantum cohomology of projective bundles over $\mathbf{P}^n$, *Trans. Amer. Math. Soc.* **350** (1998), 3615–3638, (math.AG/9607223).

[71] Bernhard Riemann, Theorie der abelschen Funktionen, *J. Reine Angew. Math.* **54** (1857), 115–155.

[72] Ziv Ran, Enumerative geometry of singular plane curves, *Invent. Math.* **97** (1989), 447–469.

[73] Yongbin Ruan and Gang Tian, A mathematical theory of quantum cohomology, *J. Diff. Geom.* **42** (1995), 259–367.

[74] Hermann Schubert, *Kalkül der abzählenden Geometrie*, Teubner, Leipzig, 1879; reprint, Springer-Verlag, Berlin, 1979.

[75] Richard P. Stanley, *Enumerative Combinatorics, Vol. 1, 2.* Cambridge University Press, Cambridge, 1997, 1999.

[76] Jakob Steiner, Allgemeine Eigenschaften der algebraischen Curven, *Berichte der Akad. Wiss. zu Berlin* (1848), 310–315; *J. Reine Angew. Math.* **47** (1853), 1–6; "Ges. Werke," herausg. von K. Weierstrass, Berlin 1882, **2**, 495–500.

[77] Stein Arild Strømme, *On parametrized rational curves in Grassmann varieties*, In: *Space Curves (Rocca di Papa, 1985)*, Vol. 1266, Lecture Notes in Math., Springer-Verlag, Berlin, 1987, pp. 251–272.

[78] Stein Arild Strømme, Elementary introduction to representable functors and Hilbert schemes, In: *Parameter Spaces (Warsaw, 1994)*, Vol. 36, Banach Center Publ., Polish Acad. Sci., Warsaw, 1996, pp. 179–198. Available at http://www.impan.gov.pl/~pragacz/parameter_spaces.htm.

[79] Jesper Funch Thomsen, Irreducibility of $\overline{M}_{0,n}(G/P, \beta)$, *Internat. J. Math.* **9** (1998), 367–376, (alg-geom/9707005).

[80] Cumrun Vafa, Topological mirrors and quantum rings, In: *Essays on Mirror Manifolds*, pp. 96–119, Internat. Press, Hong Kong, 1992.

[81] Israel Vainsencher, *Classes Características em Geometria Algébrica*, Instituto de Matemática Pura e Aplicada, Rio de Janeiro, 1985.

[82] Israel Vainsencher, Enumeration of n-fold tangent hyperplanes to a surface, *J. Alg. Geom.* **4** (1995), 503–526, (alg-geom/9312012).

[83] Ravi Vakil, The characteristic numbers of quartic plane curves, *Canad. J. Math* **51** (1999), 1089–1120, (math.AG/9812018).

[84] Ravi Vakil, Recursions for characteristic numbers of genus one plane curves, *Ark. Mat.* **39** (2001), 157–180.

[85] Herbert S. Wilf, *generatingfunctionology*, Academic Press Inc., Boston, MA, 1990. Available at http://www.math.upenn.edu/~wilf/gfology.pdf.

[86] Edward Witten, Two-dimensional gravity and intersection theory on moduli space, *Surveys in Diff. Geom.* **1** (1991), 243–310.

[87] Hieronymus G. Zeuthen, Almindelige Egenskaber ved Systemer af plane Kurver, *Vidensk. Selsk. Skr.*, 5 Række, naturvidenskabelig og mathematisk Afd., Nr. 10, B. IV, 286–393, København, 1873.

Index

GPSR Compliance
The European Union's (EU) General Product Safety Regulation (GPSR) is a set of rules that requires consumer products to be safe and our obligations to ensure this.

If you have any concerns about our products, you can contact us on

ProductSafety@springernature.com

In case Publisher is established outside the EU, the EU authorized representative is:

Springer Nature Customer Service Center GmbH
Europaplatz 3
69115 Heidelberg, Germany

www.ingramcontent.com/pod-product-compliance
Lightning Source LLC
Chambersburg PA
CBHW071709290626
47387CB00004B/370

* 9 7 8 0 8 1 7 6 4 4 5 6 7 *